Nanomedicines and Nanoproducts

Applications, Disposition, and Toxicology in the Human Body

Nanomedicines and Nanoproducts

Applications, Disposition, and Toxicology in the Human Body

Eiki Igarashi

CRC Press
Taylor & Francis Group
Boca Raton London New York

CRC Press is an imprint of the
Taylor & Francis Group, an **informa** business

CRC Press
Taylor & Francis Group
6000 Broken Sound Parkway NW, Suite 300
Boca Raton, FL 33487-2742

First issued in paperback 2017

ISBN-13: 978-1-4987-0662-9 (hbk)
ISBN-13: 978-1-138-74916-0 (pbk)

Library of Congress Cataloging-in-Publication Data

Igarashi, Eiki, author.
   Nanomedicines and nanoproducts : applications, disposition, and toxicology in the human body / Eiki Igarashi.
     p. ; cm.
   Includes bibliographical references and index.
   ISBN 978-1-4987-0662-9 (hardcover : alk. paper)
   I. Title.
   [DNLM: 1. Nanomedicine. 2. Drug Delivery Systems. 3. Nanoparticles. QT 36.5]

RS201.N35
615'.6--dc23   2014042940

Visit the Taylor & Francis Web site at
http://www.taylorandfrancis.com

and the CRC Press Web site at
http://www.crcpress.com

Contents

Preface

Long before the worldwide adoption of nanotechnology, and indeed from ancient until recent times, nanoscale substances have been utilized for countless purposes without undue attention to their size. Until the present-century nanotechnology-based research and development, basic and applied scientists had not intentionally investigated whether nanoscale substances possess unique properties that render them appropriate for special applications. Indeed, the basic sciences, with the possible exception of microscopy, were until recently developed without the knowledge of the nanoscale world. Accordingly, we now have the opportunity to reconsider evidence accumulated in the past in terms of nanoscale dimensions and attributes across various scientific fields, including physical chemistry, engineering, and biomedical research.

In principle, nanotechnology-based research began with chemical engineers and pharmacists, the bulk of whose products exhibit nanoscale dimensions. However, the range of nanoproducts now extends into two broadly defined areas: (1) the integrated sciences (i.e., the combined study of chemical engineering, pharmaceutical science, pharmacokinetic and toxicokinetic science, analytical science, microanatomy, and biomedical science) and (2) selected scientific specializations (i.e., analytical science, pharmacokinetic science, and toxicological science, each independently considering the distinctive characteristics of nanoproducts from a particular vantage point). Therefore, the design, manufacture, and practical application of nanoproducts for truly beneficial purposes will require the cooperative effort of a number of dedicated investigators from divergent scientific backgrounds, all working together as a professional team.

This book is dedicated to the discussion of nanoproducts and nanomedicines and is composed of seven chapters: (1) a synopsis of the nanoscale world; (2) an overview of the disposition of nanoproducts within the body, with a special focus on (3) the respiratory and olfactory routes of nanoproduct administration; (4) buccal exposure and the ingestion of nanoproducts; (5) the integumentary system (formed by the skin, hair, nails, and associated glands) and the ocular route; (6) the systemic route; and (7) toxicology as it relates to the nanoscale world. Chapter 1 summarizes key points for the entry of nanoproducts, nanomedicines, and other nanoscale structures into the body. Chapter 2 reviews the bodily administration and subsequent disposition of nanoproducts via various exposure routes, including the respiratory, olfactory, buccal, gastrointestinal tract, skin, follicular unit, ocular, and intravenous routes. Chapters 3 through 6 feature a detailed review of nanoproduct administration and distribution via the various routes, in addition to a discussion of practical nanoproduct and nanomedicine applications, disposition, and toxicology. The intravenous route of administration for the therapeutic management of ocular disease is also described in association with the integumentary system and the ocular route. Chapter 7 summarizes various toxicological principles and the testing of nanoproducts.

When we design a new nanoproduct for practical application, we must take into account the interactions between the nanoproduct in question and the internal environment of the body. Understanding such interactions necessitates a consideration of the assorted characteristics of nanoproducts and the nanoscale world, such as the physicochemical parameters and polydispersity of nanoproducts, their disposition within the body, and the penetration barriers set up by the body itself (Chapter 1).

The practical application of nanoproducts covers issues such as their engineering and design, encompassing the nanomaterials involved and their chemical composition, in addition to the utilization of nanoproducts as nanocosmetics, nanofoods, and nanomedicines. A new nanoproduct can be administered or exposed to human subjects via various routes, as described earlier (e.g., ocular, inhalation, oral, integumentary system, and intravenous routes), depending on the underlying reason for its use. Accordingly, the administration and disposition of nanoproducts within the body are compared with those of conventional products in terms of kinetic phases (liberation, absorption distribution, metabolism, and excretion) and parameters (bioavailability, volume of distribution, and elimination) (Chapter 2).

The respiratory system and the olfactory bulb are exposed to nanoscale substances on a daily basis (e.g., environmental chemicals or deliberately administered therapeutic agents). The practical application, disposition, and toxicology of nanoproducts absorbed via the respiratory and olfactory routes are therefore discussed for various nanomaterials, nanocosmetics, nanomedicines, and other consumer products, both organic and inorganic, in which nanotechnology has been used to enhance performance (Chapter 3).

Food is degraded into nanoscale compounds during the process of digestion, and nanoscale substances are themselves consumed by ingestion and inhalation. Chapter 4 explores the practical application, disposition, and toxicology of nanoproducts (i.e., nanomedicines, nanofoods, nanonutriceuticals, and nanodevices employed for food packaging) that are administered via the buccal route and by ingestion.

The integumentary system and the ocular organ also undergo daily exposure to nanoscale environmental substances, in the same manner as the respiratory system and the olfactory bulb. Moreover, the ocular route is subject to the effects of sunlight. Chapter 5 addresses the practical application, disposition, and toxicology of organic and inorganic nanomaterials, nanocosmetics, and nanomedicines that are exposed to the body via the integumentary system and the ocular route.

Naturally, nanoproducts lack the capacity to move throughout the body in and of themselves. However, bodily components (extracellular matrix, basal lamina, fluids, etc.) actively participate in their translocation, just as they do in the translocation of metastatic cells during pathogenic processes, the migration of normal cells during embryogenesis and morphogenesis, and the entry of foreign particles into the body, such as viruses or bacteria. The two key elements of nanoparticle translocation within the body comprise the physicochemical properties of the nanoparticles (size, shape, charge, and rigidity) and the physiological properties of the anatomical ultrastructures composing the endothelium, epithelium, and basal lamina (again, size, shape, charge, and rigidity). Chapter 6 discusses the administration and distribution of nanoproducts, especially nanomedicines, via the systemic route.

Finally, the toxicology of nanotechnology-based products must be reconsidered in terms of a specific "nano-unique" concept that encompasses toxicology principles on the nanoscale and descriptive toxicity testing of nanoproducts (Chapter 7).

As a final note, the author of this book, Eiki Igarashi, first engaged in the development of novel nanoproducts a decade ago. Through the nonclinical development of the first biodegradable block polymer in a nanobioventure company, the author has long been fascinated by nanotechnology and the nanoscale world. In writing this book, the author obtained recent information and nanoproduct reviews from several senior scientists in the fields of anatomy, drug delivery, and other basic sciences; these scientists are gratefully acknowledged later. The author surveyed the reviews and publications from these and other internationally recognized scientists and attempted to provide important information from a neutral standpoint incorporating both academic and industrial science.

This book addresses the scope of practical nanoparticle applications in academic research as well as industrial investigation, where "practical" is defined as advantageous in all aspects of nanoproduct disposition, efficacy, and toxicology. The goal of this book is to introduce potentially beneficial applications and exciting topics regarding nanoproducts to a wide variety of professional and general readers. The nanoproducts are classified herein as nanomaterials, nanofoods, nanocosmetics, and nanomedicines.

Acknowledgments

This book benefited from the nano-risk survey team in the Nanosystem Research Institute (NRI), National Institute of Advanced Industrial Science and Technology (AIST; Tsukuba, Japan).

Dr. Reiji Mezaki (deceased) and Dr. Shuji Abe provided useful comments and suggestions during the early stages of the writing of this text. Dr. Yutaka Hayashi provided partial financial support.

I acknowledge with gratitude the following reviewers and contributors:

Yoshie Maitani, professor emeritus, Hoshi University (Tokyo, Japan), and guest professor, Keio University (Tokyo, Japan).

Kazuhiro Kagami, manager of the Technical Development Division, Activus Pharma Co., Ltd. (Chiba, Japan).

Author

Eiki Igarashi is the CEO of Nanosion Co., Ltd., a Japanese nanobioventure company located in Tokyo, Japan, and a guest investigator at the National Institute of Advanced Industrial Science and Technology (AIST), also located in Tokyo, Japan. His first research endeavors, beginning in 1986, were conducted with the developmental and reproductive toxicology group led by Professor Mineo Yasuda in the Department of Anatomy at the School of Medicine, Hiroshima University (Hiroshima, Japan). Thereafter, his research interests moved into industrial exploration with the toxicology section of the Japanese branch of Dow Chemical Company in Hirakata, Japan. Concomitantly, Dr. Igarashi continued to conduct academic research in collaboration with the developmental and reproductive toxicology group led by Professor Kohei Shiota in the Department of Anatomy at the School of Medicine, Kyoto University (Kyoto, Japan). In addition, he was the project manager for a drug development team involved in nonclinical, clinical, and regulatory affairs in the pharmaceutical company Marion Merrell Dow Co., Ltd. and Hoechst Marion Roussel Co., Ltd.

Dr. Igarashi returned to Hiroshima University and earned a doctoral degree in 1998. Since 2004, he has been fascinated by nanomedicine and nanotechnology-based research, and he has dedicated himself to joint undertakings with another nanobioventure company, NanoCarrier Co., Ltd., to develop the first self-degradable, polymeric nanomedicines in the world. Dr. Igarashi also designed a translation research program for the nonclinical development of novel nanomedicines in collaboration with a limited number of professional scientists, predominantly chemical engineers. To broaden his knowledge of nanotechnology and his ability to assess nanoproduct safety and efficacy, the author researched and published a review article in 2008 focusing on the nonclinical development of polymeric nanomedicines; this endeavor involved interviews with polymer engineers and an extensive study of the literature.

In 2008, Dr. Igarashi established Nanosion Co., Ltd., a new nanobioventure company targeting the diagnosis, treatment, and palliative care of oncology diseases. Since 2009, Dr. Igarashi has worked together with the Nanosystem Research Institute of AIST to collect and review medical information regarding the practical application, toxicology, and pharmacokinetics of nanoproducts and nanomaterials, ranging from metallic nanoparticles and carbon nanotubes to nanomedicines, nanofoods, nanocosmeceuticals, and nanoconsumer products.

List of Abbreviations

AAL	*Aleuria aurantia* fungus–derived lectin
AFM	Atomic force microscopy
AMD	Age-related macular degeneration
AUC	Area under the curve
BALT	Bronchus-associated lymphoid tissue
CaSR	Calcium-sensing receptor
CCM	Cutaneous malignant melanoma
C_{max}	Peak drug concentration
ConA	Concanavalin A
DLS	Dynamic light scattering
DSPC	L-α-distearoyl-phosphatidylcholine
EC cell	Enterochromaffin cell
ED_{50}	Median effective dose
E-NTPDase	Ectonucleoside triphosphate diphosphohydrolase
EPC	Egg phosphatidylcholine
EPR	Enhanced permeability and retention
FAE	Follicle-associated endothelium
FDA	Food and Drug Administration
GALT	Gut-associated lymphoid tissue
GFR	Glomerular filtration rate
GI	Gastrointestinal
GLAST	Glutamate/aspartate transporter
GPCR	G protein–coupled receptor
GRAS	Generally recognized as safe
HMW	High molecular weight
H/P	Height to particle size ratio
ICP-MS	Inductively coupled plasma mass spectrometry
ICRP	International Commission of Radiological Protection
Ig	Immunoglobulin
^{192}Ir	Iridium isomer
K_p	Permeability constant
LADME	Liberation, absorption, distribution, metabolism, and excretion
LD_{50}	Median lethal dose
LMW	Low molecular weight
LTB	*E. coli* heat-labile toxin
MALT	Mucosa-associated lymphoid tissue
M cell	Microfold cell
MCG	Membrane-coating granule
MCNT	Multiwalled carbon nanotube
MFD	Maximal feasible dose

MLogP	Moriguchi log P
MMAD	Mass median aerodynamic diameter
MRI	Magnetic resonance imaging
NPC	Nuclear pore complex
OR	Olfactory receptor
PEG	Polyethylene glycol
PF-UV	UVA protection factor
PKD1L3	Polycystic kidney disease 1-like 3
PKD2L1	Polycystic kidney disease 2-like 1
PLGA	Poly(lactic-co-glycolic acid)
PV-1	Plasmalemmal vesicle–associated protein-1
SEM	Scanning electron microscopy
SG	Sitosteryl glucoside
SNP	Silver nanoparticle
SPF	Sun protection factor
SSTR	Somatostatin receptor
T1R	Taste-1 receptor
T2R	Taste-2 receptor
^{99m}Tc	Technetium isomer
TEM	Transmission electron microscopy
TL	Tomato lectin
TRPM5	Transient receptor potential cation channel subfamily M member 5
UV	Ultraviolet
VIR	Vomeronasal type I receptor
VNO	Vomeronasal organ
WGA	Wheat germ agglutinin

1

Nanoscale World

Nanotechnology refers to the fabrication and application of nanoscale substances, devices, and systems generated by controlling the structure at the atomic, molecular, and supramolecular level. The National Nanotechnology Initiative (a U.S. federal program to coordinate basic molecular nanotechnological research and development) refers to nanotechnology as "the understanding and control of matter at dimensions of roughly 1–100 nm, where unique phenomena enable novel applications."

The physicochemical properties of gold particles undergo pronounced changes at nanoscale dimensions, while those of polymeric micelles and liposomes do not. In fact, many substances fail to exhibit unique properties at nanoscale dimensions, while others exhibit drastic alterations. The distinctive nanoscale attributes of each substance must therefore be separately defined. Furthermore, when we consider how nanoproducts are affected by the body, the interactions between the particular nanoproduct and the body's internal environment can also be defined according to the nanoscale attributes of living tissues and cells.

In this chapter, I present an overview of the nanoscale world by summarizing the scope of currently available nanoproducts and their physicochemical properties, disposition within the body, and polydispersity. I also discuss the penetration barriers set up by the body itself.

In this book, I use the following six terms to describe the nanoscale world: nanomedicines, nanocosmetics, nanofoods, nanomaterials, nanoparticles, and nanoproducts. The first three terms, nanomedicines, nanocosmetics, and nanofoods, refer to specific types of nanoproducts that are linked with particular applications. For example, nanomedicines are generated for utilization in therapeutic and diagnostic applications in humans and domestic animals, although their use is still associated with safety concerns that must be addressed. Drugs or biologically active substances are referred to as a marketed medicine when they are already available on the market, as a candidate compound when they are at the early experimental stage, and as a project compound or project medicine when they are in the nonclinical or clinical development stage (Spilker, 1994). However, the term nanomedicine is used in this book for nanoscale therapeutic and diagnostic compounds at all stages of research and development.

The term nanomaterial does not necessarily refer to a specific industrial or practical application or to the generation of a specific nanoproduct. Rather, this term describes any material, either naturally occurring or engineered, with an average particle size of 1–100 nm. The exposure of humans to nanomaterials can occur following the deliberate administration of nanomedicines, nanocosmetics, or nanofoods, following unintentional exposure to nanoscale substances such as asbestos in the house, or after the release of nanoparticles from, for instance, nonbiodegradable materials used in food packaging.

In this book, I use the term nanoproducts in almost the same way as I use the term nanoparticles. Indeed, nanoproducts and nanoparticles can be used as integrated or near-interchangeable terms to refer to nanomedicines, nanocosmetics, nanofoods, and nanomaterials.

1.1 Scope of Nanoproducts

The scope of nanoproducts, characterized by whether or not their manufacture incorporates man-made materials and industrial technology, includes (1) naturally occurring and (2) industrially engineered nanoproducts. Naturally occurring nanoproducts comprise nanoparticles generated from industrial or naturally occurring, large-scale materials by global environmental forces, as well as nanoparticles generated from low-molecular-weight (LMW) compounds by the body's own internal machinery. On the other hand, industrially engineered nanoproducts comprise stabile or biodegradable, man-made nanoproducts specifically generated for a variety of practical applications, such as nanocosmetics for promoting health and beauty of the skin, hair, etc.; nanofoods produced via nanotechnology to enhance the nutritional value and taste (sweet, salty, bitter, sour, and umami), color, flavor, and texture of food and to improve the retention, stability, and preservation of animal and plant food sources; and nanomedicines with enhanced efficacy and safety for medical therapy and diagnosis. Finally, the range of raw materials used to generate nanoproducts includes both inorganic compounds (e.g., gold and silver) and organic constituents (e.g., amino acids and lipids).

Nanoparticles are characterized by distinctive, size-dependent physicochemical properties (Auffan et al., 2009), which dramatically affect their suitability for practical applications, disposition, and toxicology (Igarashi, 2008; Stern and McNeil, 2008; McNeil, 2009). The five predominant physicochemical attributes of nanoparticles are shape and length, chemical composition, shape-changing capacity (i.e., shape changes in response to temperature or other environmental factors), stability, and capacity for adsorption.

The first attribute refers to the shape and length of the intact nanoparticle at the date of manufacture or as it naturally exists in the environment, before undergoing any modification in the body. The length of a spherical nanoparticle is expressed as the diameter, while the length of a nanoparticle having various heights and widths (e.g., a nonuniform fiber) is expressed as

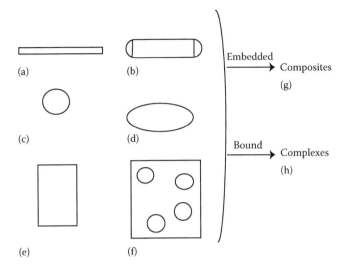

FIGURE 1.1
Classification of nanoproducts by nanoparticle shape and form (single versus mixed): (a) nanofiber without ventricles; (b) cylindrical nanotube; (c) spherical nanoparticle; (d) ellipsoidal nanoparticle; (e) nanoplate; (f) nanoplate with pores; (g) composites as embedded by (a) through (f); and (h) complexes as bound by (a) through (f). (Modified from ISO 2008. *Tech. Spec,* ISO/TS 27687 [Corrected version in February 1, 2009].)

the longest and shortest lengths in three dimensions. Shape and length are factors that affect both the disposition of the nanoparticle and its interaction with other nanoparticles and the environment. Nanoproducts for industrial and clinical applications can be classified into single or primary forms, such as nanofibers, cylindrical nanotubes, spherical particles, ellipsoidal particles, and nanoplates, as well as mixed forms, such as embedded composites and bound complexes (Figure 1.1).

Most nanoparticles utilized for the production of nanomedicines, nanofoods, and nanocosmetics are in the form of liposomes, polymer micelles, nanoemulsions, and nanocrystals (e.g., NanoCrystal®, manufactured by Elan Pharma International Ltd., Dublin, Ireland). These nanoparticle forms have relatively simple nanostructures (Figure 1.2). Others, including multilamellar liposomes and double nanoemulsions, exhibit more complicated, nesting nanostructures.

Nanoparticles can be made from a number of chemical components, including organic carbon, inorganic metals, polymers, lipids, and peptides. Chemical composition profoundly affects the disposition of a nanoproduct within the body, in addition to its biodegradation into smaller compounds or fragments.

Qualitative and quantitative shape changes result from the elasticity or surface stiffness of the nanoproduct. Shape changes attributed to the

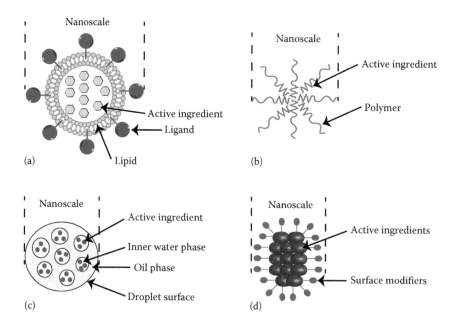

FIGURE 1.2
Classification of nanoproducts by nanostructure: (a) liposome; (b) polymer micelle; (c) nano-emulsion; and (d) nanocrystal.

flexibility/strength/elasticity of nanoparticles are discussed as rigidity in Chapters 5 and 6.

Nanoparticle stability is defined herein as the absence of any change in nanoparticle shape or size due to degradation and is influenced by nanomaterial homogeneity, the presence versus the absence of impurities, and the capacity to withstand stress. Nanoparticle stability is also referred to as "releasability" or degradability and contributes to the residual properties of an administered nanoproduct in the body resulting from delayed breakdown, liberation, and secretion.

The capacity for adsorption is attributed to nanoparticle charge and van der Waals forces. The charge or zeta potential can be neutral, cationic, or anionic, depending on the net surface charge of the nanoparticle. The charge affects absorption to cells and tissues and, therefore, the bodily disposition of the nanoparticle.

1.2 Disposition of Nanoproducts and Nanoparticles

The disposition of a nanoproduct is affected by its physicochemical properties and interactions with the tissues and cells of the body. The main physicochemical properties influencing nanoparticle disposition are size, shape, and charge, whereas the main tissue factors affecting nanoparticle disposition are the ultrastructural features of the epithelium, which forms the surface of the skin, intestines, and trachea; the mesothelium, which forms the membranous lining of the pleural and abdominal cavities and the surface of the serosa; and the endothelium, which forms the endocardium, the reticular connective tissue of the liver, and the capillaries, sinusoids, and lymph vessels. Lastly, the main cellular factors affecting nanoparticle disposition are the size, shape, and charge of molecules (cadherins, laminins, collagens, fibronectin, tight junction proteins, and so on) comprising the extracellular matrix and the epithelial, mesothelial, and endothelial basement membranes. Furthermore, the concentration or dose of the nanoparticle in question can affect its disposition, depending on the route of exposure/administration.

1.3 Polydispersity of Nanoparticles and Penetration Barriers in the Body

Nanoparticles and penetration barriers in the body show polydispersity in size. A conventionally engineered chemical compound is identified by its structural formula or molecular weight and its global categorization by the Chemical Abstract Service. On the other hand, the majority of nanoparticles produced by nanotechnological engineering have similar nanostructures, but individual particles differ in molecular weight, size, and shape. Nanoparticles also show polydispersity in bulk, even when the same engineering or manufacturing technique is used for their production. The size distribution of nanoparticles is usually expressed by a normal distribution, whereas polydispersity includes single-peak and multiple-peak cases. Thus, nanoparticles in bulk include particles with a variety of sizes, and the distribution of particle sizes often follows a normal probability distribution.

Physiological barriers to the liberation, absorption, distribution, and excretion of nanoparticles include small and large polydisperse pores, such as those found in the glomerular capillary endothelial cells of the kidney (Maul, 1971; Bearer et al., 1985) and the sinusoid endothelial cells of the liver (Muto, 1975; Ishimura et al., 1978). The polydispersity of endothelial cell pores is also characterized by single- and multiple-peak cases. In this regard, the glomerular endothelial cell pore exhibits a single peak corresponding to a diameter of approximately 5 nm, while the liver sinusoidal endothelial cell fenestrae yield multiple peaks of 100, 100–500, and 500 nm or more.

1.4 Physicochemical Properties of Nanoparticles

The physicochemical properties of nanoparticles are exploited to maximize their advantages for industrial applications and to affect their disposition for biomedical applications. Examples of nanoparticle properties relevant to industrial applications are the size-dependent attributes of gold nanoparticles and the alignment-dependent attributes of armchair-, zigzag- and chiral-type carbon nanotubes (Hamada et al., 1992; Saito et al., 1992; Dresselhaus et al., 1995). As discussed earlier, the main properties that affect nanoparticle disposition are size, shape, surface charge, and surface chemistry, while other properties include surface area, solubility according to hydrophilicity or hydrophobicity, biodegradability, density/rigidity, and the polydispersity index (Oberdorster, 2007; Sager et al., 2007; Unfried et al., 2007; Aggarwal et al., 2009; Aillon et al., 2009; Zolnik and Sadrieh, 2009).

The electronic properties of nanoparticles are important for their self-assembly via chemical bonding or physical adsorption. Moreover, electronic properties are involved in the transition from a small to a large particle size via aggregation of two or more nanoparticles chemically bound together or via agglomeration of aggregated nanoparticles and/or nanoparticles joined together by physical adsorption or interaction.

The physicochemical properties of nanoparticles undergo various changes depending on whether the environment is static or dynamic. A static environment is appropriate for the manufacture of uniform nanoparticles in a powder form, whereas an aerodynamic, hydrodynamic, or combined aerodynamic/hydrodynamic environment can substantially modify nanoparticle size, shape, and polydispersity.

1.4.1 Measurement of Nanoparticles

Dynamic light scattering (DLS) is employed to analyze the size of nanoparticles dispersed in liquid, enabling statistically constant quantitative results. Nanoparticles measured by DLS are usually larger than those measured by transmission electron microscopy (TEM). This is because the DLS-assessed size is influenced by Brownian motion and depends on the ambient temperature, the dynamic radius of the nanoparticle, and the extent of nanoparticle agglomeration triggered by a static environment via the occurrence of confliction (Gebauer and Treuel, 2011). However, recent measurement techniques enable more accurate determination of nanoparticle size than DLS affords. For example, in an analysis of colloidal nanoparticles in liquid, the diameter measured by the newly developed Brownian motion nanoparticle sizer was 36 nm, intermediate between the 32 nm diameter detected by TEM and the 42 nm diameter detected by DLS with correction for Brownian motion defects (Gebauer et al., 2012).

1.4.2 Nanoparticle Size, Shape, Density, and Rigidity

The permeability of a nanoparticle is influenced by its size, shape, density, and rigidity together with the size, shape, density, and rigidity of the permeability barriers within the body. Nanoparticle size measured by DLS is indirectly assessed by Brownian motion, under the assumption that the particle assumes a spherical shape. Therefore, the size is estimated by the average diameter and standard division in three dimensions, even if the shape is in fact an elongated form, an ellipse, or an infinite form rather than a sphere. When nanoparticles circulate in the body and pass through permeability barriers, the critical limiting diameter is the least diameter in three dimensions.

Nanoparticle shape changes within the confines of the body are contingent on the density and rigidity of the particle. For example, liposomes are classified as soft or hard by rigidity, where soft liposomes are less rigid than hard liposomes and can change their shape more readily. Hard liposomes (~150 nm in diameter) require a fairly long time to distribute in the Disse space of the liver because the size distribution of hepatic endothelial cell pores is such that ~70% of the pores have a diameter of ≤100 nm and ~30% have a diameter of >100 nm. Soft-type liposomes, by comparison, are rapidly distributed in the liver due to their shape-changing capacity, permitting passage through the smaller endothelial cell pores of 100 nm or less (Maitani, 1996).

Coated nanoparticles generally exhibit a high-density inner core together with a low-density outer core (Chapter 7). Therefore, coated nanoparticles with a spherical shape after manufacture must frequently undergo modification into an elliptical shape in the body because not all permeability barriers are spherical. Notably, glomerular capillary pores are fan shaped (Maul, 1971; Bearer et al., 1985), while the sinusoidal fenestrae are spherical (Muto, 1975; Ishimura et al., 1978).

1.4.3 Biodegradation of Nanoparticles and Endocytosis

Nanoparticle biodegradability can usually be introduced via specific engineering techniques, depending on the desired application and the chemical composition of the particle. Nanoproducts with rigidity as an advantage have limited or no biodegradability. However, it is essential that nanomedicines and nanocosmetics be biodegradable to avoid long-term accumulation in the body. Because nanomedicines are generally administered to the patient on a repeated basis, they must also be stealthy, that is, unrecognizable as a foreign body by the patient's immune system. Another consideration regarding the biodegradability and stealth of nanomedicines is that accumulated intermediates or degraded fragments can themselves be active and endocytosed by cells.

1.4.4 Nanomedicines and Electrostatic Charge

The desired electrostatic properties of nanomedicines for medical applications frequently differ from those of nanomaterials for industrial applications. Nanomedicines for use as drugs are designed to minimize harmful effects to the body by, for example, neutralization of surface charge, optimization of dispersion/degradation, and promotion of solubility. By contrast, many nanoproducts for use in industrial applications are designed such that their properties (e.g., electrical conductivity) show long-term stability. Electrostatic charge or electrical conductivity is common in nanomaterials and may enhance nanoparticle/cell interactions by increasing particle adsorption to the cell surface. However, charged nanomaterials may also be subject to chemical aggregation or physical agglomeration during manufacture, which is not typically conducive to their use as drugs. Even though intact nanomaterials can be neutralized in advance of aggregation or agglomeration by chemical modification, the same nanomaterials can revert to their charged status after exposure to bodily elements. The advantages versus disadvantages of charged nanoproducts for medical purposes must therefore be considered on a case-by-case basis.

1.4.5 Nanoparticle Surface Area and Cell Encounters

The interaction of a nanoparticle with a cellular membrane can be physical or chemical. Physical interactions depend on the cationic, neutral, or anionic charge of the nanoparticle, as well as the expression of cation, neutral amino acid, or anion transporters by the target cell. On the other hand, chemical interactions depend on the presence of specific cellular ligands on the surface of coated nanoparticles and the expression of corresponding receptors on the target cells. Because the surface area of a conventional substance is smaller than that of a nanoparticle, the statistical probability of a cell encounter by a conventional substance versus a nanoparticle is relatively low. Thus, if a high-affinity interaction with a cell surface receptor or a plasma membrane transporter is required for an optimal biological effect, the enhanced density of reactive positions on the nanoparticle surface will likely prove advantageous.

1.5 History of Nanotechnology and Nanomedicine

Silver colloid nanoparticles of 7–9 nm have been in use for 120 years (Nowack et al., 2011), and nanoscale liposomes were first produced and described by Bangham in 1965 (Bangham, 1993). In 1974, nanotechnology was defined by Taniguchi as "the process of separation, consolidation and deformation

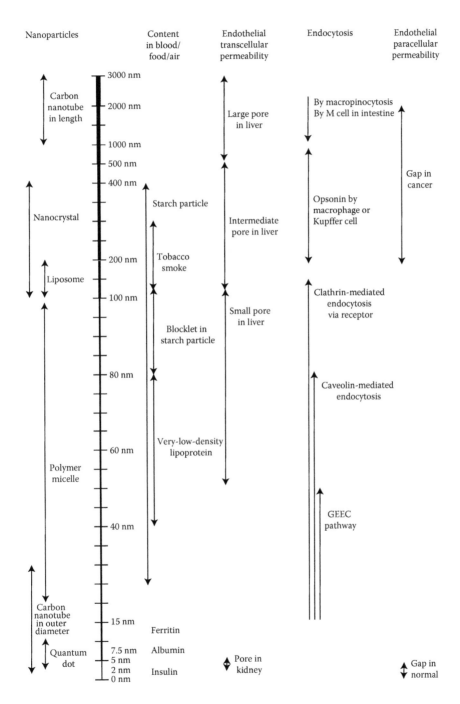

FIGURE 1.3
The nanoscale world defined by nanoproducts and their interactions with the human body.

of materials by one atom or one molecule" (Taniguchi, 1994; Tibbals, 2011). Yokoyama went on to design biodegradable, injectable polymer micelles in 1990 (Yokoyama et al., 1990), and Iijima discovered carbon nanotubes in 1991 by using advanced electron microscopy techniques (Harris, 2009). Lastly, Freitas originally introduced the term "nanomedicine" in 1999 (Freitas, 2002).

1.6 Overview of the Nanoscale World

The ultrastructural dimensions of various nanoproducts in relation to the human body can be roughly estimated from the images in Figure 1.3, which shows that they have a size similar to the nanoscale features of cells, tissues, and bodily fluids (Figure 1.3). Furthermore, many compounds found in blood, food, and air are nanoscale in size, including nutrients (e.g., starch) absorbed in the intestine, physiological molecules produced by the body (e.g., very-low-density lipoprotein, insulin, ferritin, and serum albumin), and environmental contaminants found in the air (e.g., tobacco smoke) (see Chapters 3, 4, and 6). The dimensions of nanoscale liposomes and polymer micelles are optimal for injectable application (Chapter 6), while those of nanocrystal are optimal for oral administration (Chapter 4). The pathways taken by nanoproducts once inside the body are also size dependent; for example, transcellular permeability, endocytosis, and endothelial paracellular permeability are all limited by the size of the transported particles (Chapter 4). These topics are discussed in detail in the remaining chapters of this book.

References

Aggarwal, P., J. B. Hall, C. B. McLeland, M. A. Dobrovolskaia, and S. E. McNeil. 2009. Nanoparticle interaction with plasma proteins as it relates to particle biodistribution, biocompatibility and therapeutic efficacy. *Advanced Drug Delivery Reviews* 61 (6):428–437. doi: 10.1016/j.addr.2009.03.009.

Aillon, K. L., Y. M. Xie, N. El-Gendy, C. J. Berkland, and M. L. Forrest. 2009. Effects of nanomaterial physicochemical properties on in vivo toxicity. *Advanced Drug Delivery Reviews* 61 (6):457–466. doi: 10.1016/j.addr.2009.03.010.

Auffan, M., J. Rose, J. Y. Bottero, G. V. Lowry, J. P. Jolivet, and M. R. Wiesner. 2009. Towards a definition of inorganic nanoparticles from an environmental, health and safety perspective. *Nature Nanotechnology* 4 (10):634–641. doi: 10.1038/nnano.2009.242.

Bangham, A. D. 1993. Liposomes: The Babraham connection. *Chemistry and Physics of Lipids* 64 (1–3):275–285.

Bearer, E. L., L. Orci, and P. Sors. 1985. Endothelial fenestral diaphragms—A quick-freeze, deep-etch study. *Journal of Cell Biology* 100 (2):418–428. doi: 10.1083/jcb.100.2.418.

Dresselhaus, M. S., G. Dresselhaus, and R. Saito. 1995. Physics of carbon nanotubes. *Carbon* 33 (7):883–891.

Freitas, R. A., Jr. 2002. The future of nanofabrication and molecular scale devices in nanomedicine. *Studies in Health Technology and Informatics* 80:45–59.

Gebauer, J. S., M. Malissek, S. Simon, S. K. Knauer, M. Maskos, R. H. Stauber, W. Peukert, and L. Treuel. 2012. Impact of the nanoparticle-protein corona on colloidal stability and protein structure. *Langmuir* 28 (25):9673–9679. doi: 10.1021/La301104a.

Gebauer, J. S. and L. Treuel. 2011. Influence of individual ionic components on the agglomeration kinetics of silver nanoparticles. *Journal of Colloid and Interface Science* 354 (2):546–554. doi: 10.1016/j.jcis.2010.11.016.

Hamada, N., S. Sawada, and A. Oshiyama. 1992. New one-dimensional conductors—Graphitic microtubules. *Physical Review Letters* 68 (10):1579–1581.

Harris, P. J. F. 2009. *Carbon Nanotube Science*. Cambridge, U.K.: Cambridge University Press.

Igarashi, E. 2008. Factors affecting toxicity and efficacy of polymeric nanomedicines. *Toxicology and Applied Pharmacology* 229 (1):121–134. doi: 10.1016/j.taap.2008.02.007.

Ishimura, K., H. Okamoto, and H. Fujita. 1978. Freeze-etching images of capillary endothelial pores in liver, thyroid and adrenal of mouse. *Archivum Histologicum Japonicum* 41 (2):187–193.

ISO 2008. Nanotechnologies—Terminology and definitions for nano-objects—Nanoparticle, nanofibre and nanoplate *Technical Specification*. ISO/TS 27687 (corrected version in February 1, 2009).

Maitani, Y., M. Hazama, Y. Tojo, N. Shimoda, and T. Nagai. 1996. Oral administration of recombinant human erythropoietin in liposomes in rats: Influence of lipid composition and size of liposomes on bioavailability. *Journal of Pharmaceutical Sciences* 85 (4):440–445.

Maul, G. G. 1971. Structure and formation of pores in fenestrated capillaries. *Journal of Ultrastructure Research* 36:768–782.

McNeil, S. E. 2009. Nanoparticle therapeutics: A personal perspective. *Wiley Interdisciplinary Reviews—Nanomedicine and Nanobiotechnology* 1 (3):264–271. doi: 10.1002/Wnan.006.

Muto, M. 1975. Scanning electron-microscopic study on endothelial cells and Kupffer cells in rat-liver sinusoids. *Archivum Histologicum Japonicum* 37 (5):369–386.

Nowack, B., H. F. Krug, and M. Height. 2011. 120 Years of nanosilver history: Implications for policy makers. *Environmental Science & Technology* 45 (4):1177–1183. doi: 10.1021/Es103316q.

Oberdorster, G. 2007. Biokinetics and effects of nanoparticles. *Nanotechnology—Toxicological Issues and Environmental Safety* 276:15–51.

Sager, T. M., D. W. Porter, V. A. Robinson, W. G. Lindsley, D. E. Schwegler-Berry, and V. Castranova. 2007. Improved method to disperse nanoparticles for in vitro and in vivo investigation of toxicity. *Nanotoxicology* 1 (2):118–129. doi: 10.1080/17435390701381596.

Saito, R., G. Dresselhaus, and M. S. Dresselhaus. 1992. Topological defects in large fullerenes. *Chemical Physics Letters* 195 (5–6):537–542.

Spilker, B. 1994. *Multinational Pharmaceutical Companies: Principles and Practices.* New York: Lippincott Raven.

Stern, S. T. and S. E. McNeil. 2008. Nanotechnology safety concerns revisited. *Toxicological Sciences* 101 (1):4–21. doi: 10.1093/toxsci/kfm169.

Taniguchi, N. 1994. The state-of-the-art of nanotechnology for processing of ultra-precision and ultrafine products. *Precision Engineering—Journal of the American Society for Precision Engineering* 16 (1):5–24. doi: 10.1016/0141-6359(94)90014-0.

Tibbals, H. F. 2011. Medical nanotechnology and nanomedicine. In: G. L. Hornyak (ed.), *Prospectives in Nanotechnology.* Boca Raton, FL: CRC Press, Taylor & Francis Group.

Unfried, K., C. Albrecht, L. O. Klotz, A. Von Mikecz, S. Grether-Beck, and R. P. F. Schins. 2007. Cellular responses to nanoparticles: Target structures and mechanisms. *Nanotoxicology* 1 (1):52–71. doi: 10.1080/00222930701314932.

Yokoyama, M., M. Miyauchi, N. Yamada, T. Okano, Y. Sakurai, K. Kataoka, and S. Inoue. 1990. Polymer micelles as novel drug carrier—Adriamycin-conjugated poly(ethylene glycol) poly(aspartic acid) block copolymer. *Journal of Controlled Release* 11 (1–3):269–278.

Zolnik, B. S. and N. Sadrieh. 2009. Regulatory perspective on the importance of ADME assessment of nanoscale material containing drugs. *Advanced Drug Delivery Reviews* 61 (6):422–427. doi: 10.1016/j.addr.2009.03.006.

2

Overview of Nanoproduct Disposition

Nanoproducts and nanoparticles are mainly designed for four applications: (1) engineering/industrial applications, (2) cosmetic applications, (3) nutritional applications, and (4) medical or veterinary therapeutic and diagnostic applications (Figure 2.1). Humans can be occupationally or environmentally exposed to any of these nanoproducts, in particular, nanocosmetics/nanocosmeceuticals, nanofoods, and nanomedicines. The most prevalent modes of nanoproduct exposure are the ocular, oral, and dermal/transdermal routes, as well as ingestion, inhalation, and injection (intravenous, intraperitoneal, subcutaneous, intrathecal, and intramuscular). The dermal/transdermal route is particularly common for the delivery of nanocosmetics/nanocosmeceuticals, while the oral route predominates for nanofoods. All of these routes are employed for the administration of nanomedicines.

This chapter provides an overview of nanoproduct administration via the various routes described and their ensuing disposition in the body. Comparisons are also provided between nanoproducts and conventionally engineered chemical substances and drugs.

2.1 Pathways for Nanoproduct Absorption, Distribution, and Excretion

Conventionally engineered chemical substances and drugs are administered to recipients through a number of routes, including oral, buccal, ocular, intravenous, intraperitoneal, subcutaneous, intramuscular, and dermal/transdermal routes, as well as ingestion and inhalation (Figure 2.2). The route of administration/exposure affects the disposition. For example, compounds that are administered via inhalation are mainly distributed to the lung, followed by the gastrointestinal (GI) tract. They are then distributed from the GI tract to the liver via the hepatic portal vein, and thus, any compounds or breakdown products must also pass through the blood and the lymph.

Nanoproducts/nanoparticles have the same routes of administration and distribution pathways as conventional products. The assorted routes are discussed in detail in Chapters 3 through 6.

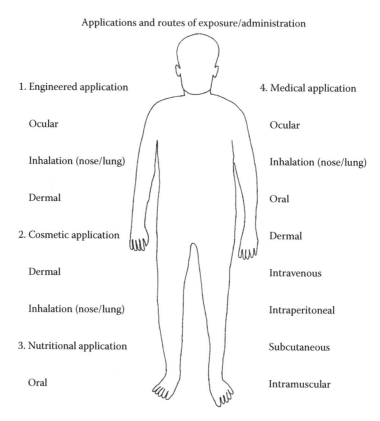

Applications and routes of exposure/administration

1. Engineered application

   Ocular

   Inhalation (nose/lung)

   Dermal

2. Cosmetic application

   Dermal

   Inhalation (nose/lung)

3. Nutritional application

   Oral

4. Medical application

   Ocular

   Inhalation (nose/lung)

   Oral

   Dermal

   Intravenous

   Intraperitoneal

   Subcutaneous

   Intramuscular

FIGURE 2.1
Routes of exposure/administration for nanoproducts designed for (1) engineering/industrial, (2) cosmetic, (3) nutritional, and (4) medical/diagnostic applications.

2.2 Liberation, Absorption, Distribution, Metabolism, and Excretion

The fate of conventional products and nanoparticles in the body after administration or exposure is classified according to five kinetic phases: liberation, absorption, distribution, metabolism, and excretion (LADME). The kinetic phases are further described by three kinetic parameters: bioavailability, volume of distribution, and elimination (Figure 2.3). The bioavailability of conventional products and nanoparticles includes absorption plus liberation or absorption alone without liberation. When orally administered, nanoparticles and nanomedicines are absorbed by the intestine and achieve systemic circulation as either intact nanomedicines (prodrugs) or active fragments

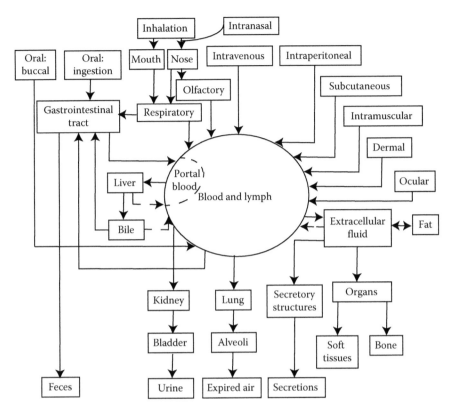

FIGURE 2.2
Routes of absorption, distribution, and excretion for inhalation, intranasal, intravenous, intra-peritoneal, subcutaneous, intramuscular, dermal, and ocular routes. (Modified from Klaassen, C.D., *Casarett and Doull's Toxicology: The Basic Science of Poisons*, 5th edn., McGraw-Hill, New York, 1995; the ocular route was added to the figure.)

thereof via pathways involving absorptive cells or via pathways involving microfold or microplicae cells (M cells).

2.3 Liberation

A biodegradable nanomedicine is itself a prodrug or an inactive drug, until the active compound is released from the delivery system in the process of liberation. Nanomedicines for oral or intravenous delivery are frequently composed of a "payload" of active ingredients incorporated into a nanocapsule, surrounded by an inactive, slowly degradable shell as the delivery platform. The platform protects the active ingredients from rapid metabolism

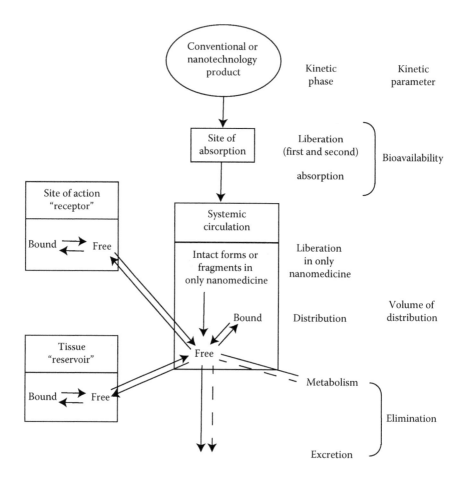

FIGURE 2.3
Fate of conventional products and nanoproducts in the body after oral administration. (Modified from Marquardt, H. et al., *Toxicology*, Academic Press, San Diego, CA, 1999; the liberation phase for nanomedicines was added to the figure.)

in the blood until they are delivered to the target cells. Conventional products can also be surrounded by inactive substances termed excipients. The biodegradability of the platform thus affects the kinetic phase of liberation.

2.4 Bioavailability

When an individual receives an intravenous injection of a conventional product, the entire dose of the product is circulated throughout the vasculature. The substance eventually enters the left ventricle of the heart via the right ventricle and the pulmonary artery. From the left ventricle, the substance in the blood is pumped via the aorta to the systemic arteries and all the organs and tissues of the body. The only exception is a drug that is strongly metabolized in the plasma or lung.

Oral and dermal routes of administration are commonly used for conventional products, necessitating that the substance penetrate the intestinal mucosal membrane or the dermal epithelium, respectively, in the process of absorption (Figure 2.4). Thus, the absorption rate and bioavailability of a conventional product depend on its capacity to pass through biological membranes.

As previously described (Chapter 1), the absorption rate of a nanoproduct depends on both the size of the nanoparticles and the size of the pores in the biological membrane through which the nanoparticles must pass, in addition to the aforementioned properties for conventional products.

2.4.1 Bioavailability of Orally Administered Products

Following the oral administration of a conventional product or a nanoproduct, the entire dose of the substance is not necessarily absorbed or bioavailable. Orally administered compounds can be absorbed from the stomach, intestine, colon, and upper portion of the rectum. They then progress to the metabolic pathway in the liver or the elimination pathway in the bile via the hepatic portal vein (Figure 2.5). While the metabolic and elimination pathways are the same for conventional products and nanoproducts, the latter has the probable advantage of an increased concentration in the portal vein. This is because nanomedicines are typically protected from degradation and/or inactivation in the intestine by encapsulation within a slowly degradable shell or association with a nanocrystal platform.

2.4.2 Bioavailability of Intravenously Injected, Biodegradable Nanomedicines

The injection time itself corresponds to the liberation phase for most conventional injectable drugs. The bioavailability after injection is regarded as 100% because these products do not experience the first pass through the liver. However, the period of 100% bioavailability is extremely short because conventional drugs are rapidly metabolized in the liver and eliminated in the kidney. Furthermore, active ingredients undergo enzymatic modifications to their chemical structure immediately after injection.

A biodegradable, injectable nanomedicine is a nanoscale formulation loaded with active, therapeutic, or diagnostic components and is exemplified by the polymer micelle. These nanoscale formulations do not show pharmacological effects in the intact form prior to drug release or degradation (Figure 2.6). Biodegradable nanomedicines are circulated throughout the vasculature following intravenous injection and then distributed to systemic tissues.

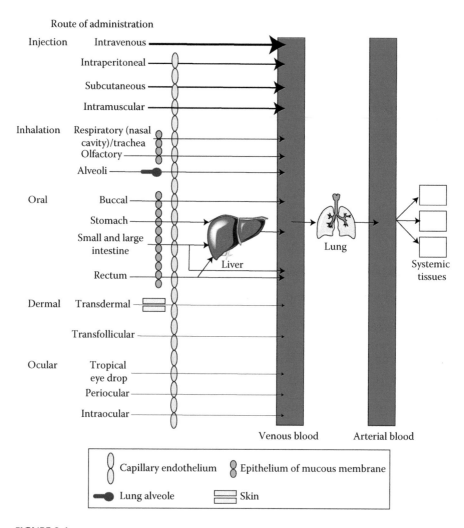

FIGURE 2.4
Schematic diagram of product bioavailability via the five predominant routes of administration (injection, inhalation, oral, dermal, and ocular routes). (Modified from Marquardt, H. et al., *Toxicology*, Academic Press, San Diego, CA, 1999; the ocular route was added to the figure.)

They can exert local effects because their active ingredients are released by degradation during circulation and accumulate in the vicinity of target cells (e.g., cancer cells). Engineered degradable nanomedicines must therefore exhibit the essential property of stealth, thereby circumventing the degradation of the inactive nanoparticle shell or surface by metabolic enzymes in the blood as well as phagocytosis by macrophages. An intact, nondegraded nanomedicine is typically unaffected by metabolism in the liver because the active ingredients are sequestered within the nanocapsule.

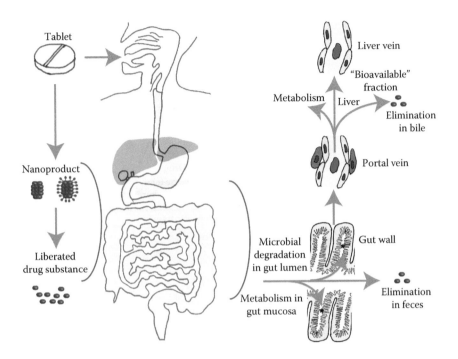

FIGURE 2.5
Bioavailability and presystemic metabolism. Only the portion of an orally administered drug that arrives unchanged in the central venous blood pool is "bioavailable." Bioavailability depends on absorption and presystemic metabolism in the intestine and the liver. A conventional drug must be liberated from inactive ingredients including excipients, whereas a nanoproduct must be liberated from the drug delivery platform or the prodrug. (Modified from Forth, W. et al., *Pharmakologie Und Toxikologie*, Spektrum Akademischer Verlag, Heidelberg, Germany, 1996; written by Fichtl et al., Allegemeine pharmakologie und toxikologie, in *Allgemeine und spezielle Pharmakologie und Toxikologie: Für Studenten der Medizin, Veterinärmedizin, Pharmazie, Chemie, Biologie sowie für Ärzte, Tieräzte und Apotheker*, W. Forth, D. Henschler, W. Rummel, and K. Starke (eds.), 6th edn., Spektrum Akademischer Verlag, Heidelberg/Berlin/Oxford, 1996, pp. 1–95.)

The timing of the release of the active therapeutic ingredients from a biodegradable nanomedicine is considered the drug liberation phase. By contrast, all of the ingredients in a biodegradable nanoparticle loaded with diagnostic agents are hypothetically active, and therefore, the timing of injection is considered to be the liberation phase. For oncology applications, a nanomedicine may be specially designed to exhibit biodegradation only in the local environment of the cancer cell. In this case, the timing of the cancer cell–induced biodegradation corresponds to the liberation phase, and thus, the intact nanomedicine, the biodegraded fragments, and the released ingredients all are considered active agents for therapy. True bioavailability is therefore assessed by the efficacy of drug release around the malignancy.

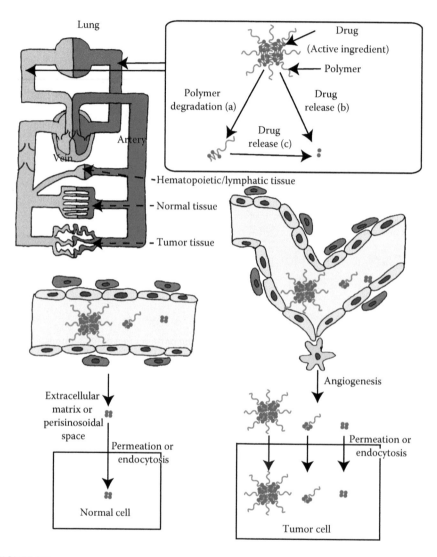

FIGURE 2.6

Schematic representation of the degradation, release, and uptake of polymeric nanomedicines into normal and tumor cells by passive and active targeting. *Note*: Arrows indicate the processes of degradation, release, and uptake. The bold arrow shows the impact of active targeting on these processes. (a) An intact polymeric nanomedicine may be endogenously degraded into degraded nanomedicines of different sizes. The drug may be released from (b) an intact or (c) a degraded polymeric nanomedicine. (Modified from Igarashi, E., *Toxicol. Appl. Pharmacol.*, 229(1), 121, 2008, doi: 10.1016/j.taap.2008.02.007.)

2.4.3 Bioavailability of Intravenously Injected, Active-Targeting Nanomedicines

The recently developed, molecularly targeted drugs refer to coupled agents composed of an active ingredient and an antibody with the capability to specifically recognize the plasma membrane of target cells (Figure 2.7a). When a molecularly targeted drug interacts with the target cell or is taken up into the cell, depending on the pharmacological mechanism, the timing of the interaction or uptake is considered to be liberation and the initiation of pharmacological action.

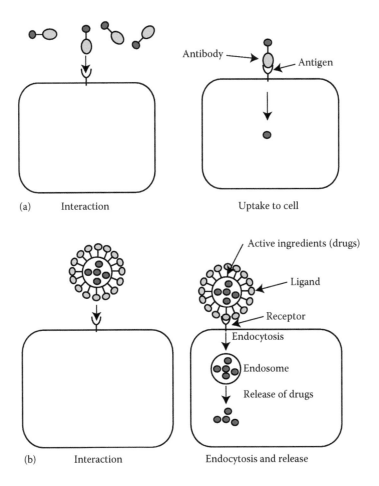

FIGURE 2.7
(a) Molecular-targeting drugs. Schematic diagram of molecularly targeted drug/cell interaction, followed by endocytosis, drug uptake into the cell, and drug release. (b) Active-targeting drugs. Schematic diagram of active-targeting nanomedicine/cell interaction, endocytosis, uptake into cell, and drug release.

On the other hand, an active-targeting nanomedicine refers to a nanocapsule or a mixed compound composed of many active ingredients loaded together into an inner shell and surrounded by an outer ligand-coated shell. Active-targeting nanomedicines are capable of specifically interacting with target cells, followed by endocytosis and the release of the active ingredients from the inner capsule (Figure 2.7b). The timing of the interaction with the plasma membrane, or more restrictively, the timing of the release of the active ingredients, is considered to be liberation for active-targeting nanomedicines.

In the case of molecularly targeted drugs, the engineering of the drug drives a somewhat limited liberation because only one active site provided by the antibody interacts with the target cell plasma membrane. In the case of active-targeting nanomedicines, the engineered design drives a highly probable liberation because numerous active sites provided by ligands on the drug surface increase the likelihood of drug/target cell interactions. Again, true bioavailability is assessed by the efficacy of drug release around or within the targeted cells.

2.5 Volume of Distribution

2.5.1 Conventional Drugs versus Nanomedicines

The volume of distribution of conventional drugs varies according to the property of the particular drug. For example, the volume of distribution of chlorpromazine is 20 L/kg, where the total volume is given as 1400 L for a human being with a body weight of 70 kg. This result suggests that chlorpromazine distributes into the plasma, extracellular matrix/fluid, and cells. However, the volume of distribution of heparin is only 0.06 L/kg, where the volume is given as 4.2 L. This suggests that heparin is mainly distributed to the plasma. Hence, heparin binding to plasma proteins lengthens its retention time in the plasma and lowers its transfer into other compartments. In this case, retention in the plasma does not necessarily contribute to beneficial pharmacological actions.

Nanomedicines show extremely small volumes of distribution immediately after injection and primarily distribute to the plasma. A disposition study of 30 nm cisplatin-loaded polymer micelles in mice demonstrated that the volume of distribution of the micelle was smaller than that of free cisplatin by a factor of 75, namely, 3 L for free cisplatin versus 0.04 L for cisplatin micelles (Uchino et al., 2005). Because nanomedicines are typically engineered with, for example, polyethylene glycol (PEG)-coated outer shells to permit rapid dissociation from plasma proteins, their small volumes of distribution are attributed to lowered blood capillary permeability rather than to augmented plasma protein binding.

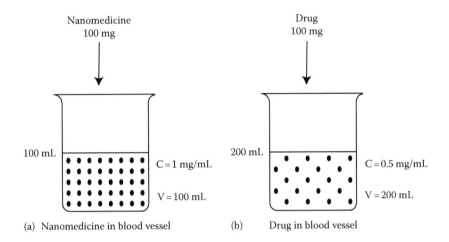

(a) Nanomedicine in blood vessel   (b)   Drug in blood vessel

FIGURE 2.8
Concept of the apparent volume of distribution within the blood vessel. In this simplified model, the body is represented as a beaker filled with water, where the water represents the plasma. (a) An amount A of nanomedicine (100 mg as drug) is dissolved in 100 mL of water, yielding a concentration B of 1 mg/mL. According to the equation $V = A/B$, the volume of distribution is calculated as 100 mg/(1 mg/mL) = 100 mL, which is equivalent to the real distribution in 3D space. (b) On the other hand, if 100 mg of drug is dissolved in 100 mL of water, and a portion of the drug is accumulated in the tissue (represented by precipitation at the bottom of the beaker), the effective concentration of the drug in water will be reduced. The total volume composed of water (plasma) and precipitation (tissue) is assumed to be 200 mL in this beaker model. (Modified from Marquardt, H. et al., *Toxicology*, Academic Press, San Diego, CA, 1999.)

A small volume of distribution heightens the concentration of a drug in the plasma. For instance, the concentration of the same amount of drug after dispersion into a small space (represented by the beaker model in Figure 2.8a) shows a higher concentration than dispersion into a large space (Figure 2.8b).

2.5.2 Special Note Regarding the Association between Increased Permeability of Capillaries in Cancer Tissue and Drug Bioavailability

Angiogenesis is prevalent in the capillary endothelium that neighbors sites of inflammation or cancer, and large pores or intercellular gaps are created in the endothelium during the angiogenic process (Figure 2.9). These pores or gaps are sufficiently sized to allow the ready passage of nanomedicines. This phenomenon, the enhanced permeability and retention (EPR) effect discovered and coined by Maeda and Matsumura (1989), asserts that compounds of a certain size (e.g., liposomes, nanoparticles, and macromolecular drugs) cannot cross the endothelial membrane of normal tissue.

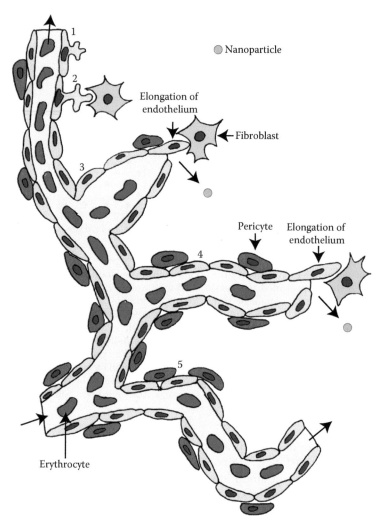

FIGURE 2.9
Angiogenesis showing gap formation in the capillary endothelium in rats. Fibroblasts, pericytes, and mast cells (not shown in the figure) participate in angiogenesis in an orderly fashion,
as described by (1)–(5): (1) Endothelial processes elongate. (2) The elongated process interacts
with a fibroblast. (3) The fibroblast pulls on the endothelium to create a gap or fenestra. (4) The
pericyte covers the open gap. (5) The irregular capillary endothelium in tumor. (Modified from
Rhodin, J.A. and Fujita, H., *J. Submicrosc. Cytol. Pathol.*, 21(1), 1, 1989.)

However, these compounds can pass through the pores or gaps in the endothelium surrounding cancer tissue, and therefore, they tend to accumulate
in tumor tissue to a greater degree than in normal tissue (Matsumura and
Maeda, 1986). This tendency is further enhanced by the fact that lymphatic
circulation does not function normally around cancer tissue.

TABLE 2.1

Diameters of Blood Vessels and Endothelial Fenestrae in Cancer Tissue

Tumor Type	Vessels, Endothelial Fenestrae, or Endothelial Gaps	Size Determined by SEM
	Vessels	Mean, 39 μm
		Range, 8–220 μm
MCa-IV mouse mammary carcinomas[a]	Endothelial fenestrae	Mean, 600 nm
		Range, 300–900 nm
		Area% per liminal surface area, 0.0008%
	Endothelial gaps	Mean, 1.7 μm
		Range, 0.3–4.7 μm
		Area% per liminal surface area, 0.1%
RIP-Tag2 pancreatic islet tumors[b]	Vessels	Mean, 8 μm
MCa-IV mouse mammary carcinomas[b]	Vessels	Mean, 45 μm
Lewis lung carcinoma[b]	Vessels	Mean, 31 μm
HCa-1 mouse hepatoma[c]	Endothelial fenestrae/gaps	Range, 380–550 nm
Shionogi male testosterone–dependent mammary carcinoma[c]	Endothelial fenestrae/gaps	Range, 200–380 nm
MCa-IV mouse mammary carcinomas[c]	Endothelial fenestrae/gaps	Range, 1200–2000 nm

Source: Adapted from Igarashi, E., *Toxicol. Appl. Pharmacol.*, 229(1), 121, 2008, doi: 10.1016/j.taap.2008.02.007

[a] Summarized from Hashizume et al. (2000).
[b] Summarized from Morikawa et al. (2002).
[c] Summarized from Hobbs et al. (1998).

The EPR effect has received worldwide recognition and is widely applied for the research and development of new cancer-targeting nanomedicines (McNeil, 2009). New "magic bullet" chemotherapeutic agents are highly anticipated based on the EPR effect and are expected to have minimal impact on normal tissue and maximal impact against tumors.

The permeability of the endothelium has been applied to develop new conventional drugs as well as nanomedicines against cancer. However, different malignancies are associated with pores or gaps of different sizes, and capillary permeability is not constant in one type or location of cancer (Table 2.1) (Igarashi, 2008). Therefore, the engineering of new therapeutic agents for oncology must consider cancer-specific endothelial pore and gap dimensions.

2.5.3 Volume of Distribution of Biodegradable Nanomedicines for Cancer Therapy

Conventional or low-molecular-weight (LMW) drugs can pass through the pores of the capillary endothelium, regardless of whether the tissue is normal or cancerous. This allows roughly equivalent drug distribution into the three main compartments of the body: the blood plasma, the extracellular matrix/fluid, and the cells (Figure 2.10a). The EPR effect predicts that an intact biodegradable nanomedicine will only distribute to the plasma in normal tissue and to all three compartments in tumors. The volume of the blood vessels is smaller than the volume of the extracellular matrix/fluid or the cells, and thus, nanomedicines have a smaller volume of distribution in normal versus cancerous tissue. Because the release of active ingredients is critical for the therapeutic actions of biodegradable nanomedicines, the volume of distribution of the released active ingredients and their retention in the tumor is of utmost importance for effective cancer therapy.

2.5.4 Volume of Distribution of Intravenously Administered Active-Targeting Nanomedicines

As described earlier, conventional drugs can cross the membrane of blood capillary cells in both normal and cancerous tissue, distributing into cells and the extracellular space as well as into the plasma. On the other hand, because intact active-targeting nanomedicines interact with the target cell, the volume of distribution of the intact form is key (Figure 2.10b). Successful active-targeting nanomedicines therefore require engineering to promote their accumulation in cancer tissue and cells at higher concentrations than those of conventional drugs.

2.5.5 Permeability of Capillaries in Normal Tissue

Many different types of capillaries are found in the body. Most conventional LMW compounds can pass through the membrane of all types of capillary endothelium, excluding drugs with attenuated permeability due to their association with plasma proteins. By contrast, nanomedicines can pass through the noncontinuous capillary wall in the liver and spleen (Figure 2.11c and d), but they cannot pass through the continuous capillary wall in normal tissues (Figure 2.11a and b).

2.6 Clearance

Clearance is the measure of the body's ability to eliminate foreign substances. The elimination of conventional drugs includes removal by excretion and reformation or conversion of chemical structures by metabolism (Figure 2.12). Elimination pathways include excretion in the urine by the kidney, secretion in the bile, excretion in the feces through the intestine, exhalation from the lungs, and partial secretion through the intestine. A small

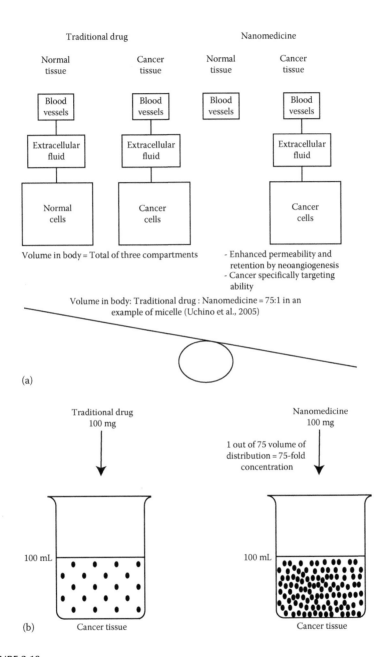

FIGURE 2.10
Volume of distribution and accumulation of conventional drugs and nanomedicines in cancer tissue. (a) The size of the box is roughly equivalent to the volume in the body and the plasma/blood vessels, which is <extracellular matrix/fluid and <cells. (b) The beaker shows drug concentrations based on an example of the volume of distribution in rats by Uchino et al. (2005). (Modified from Marquardt, H. et al., *Toxicology*, Academic Press, San Diego, CA, 1999.)

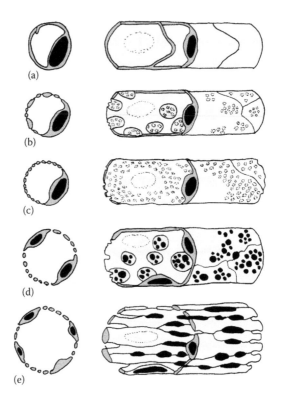

FIGURE 2.11
Schematic diagram of different types of capillary vessels. (a) Continuous capillary referred to as muscular type, (b) fenestrated or pored capillary seen in the part of endocrine, (c) fenestrated or pored capillary seen in renal glomerulus, (d) liver sinusoid referred to as hepatic type, and (e) sinus of the spleen. (Modified from Yamada, Y., *Modern Textbook of Histology, Third Edition*. Kanehara & Co., Ltd., Tokyo, 1994; Fujita, H. and Fujita, T., *Textbook of Histology, Part 2*, 3rd edn., Igaku-Shoin Ltd., Tokyo, Japan, 1992.)

portion of the drug is eliminated in the saliva, tears, and sweat, in addition to the milk of lactating mothers. Conventional compounds and nanomedicines are eliminated via the same pathways, where conventional compounds are tracked as the drug itself, and nanomedicines are tracked as the released active ingredients or degraded fragments.

2.6.1 Renal Clearance

Renal clearance occurs via three mechanisms: glomerular filtration, tubular reabsorption, and tubular secretion (Figure 2.13). The renal plasma flow is approximately 600 mL/min, whereas the normal glomerular filtration rate (GFR) of conventional LMW drugs is 120 mL/min for typical glomerular capillary pores with a size of ~5 nm (see Chapter 6). However, the GFR of >5 nm nanomedicines is 0 mL/min (Figure 2.13a).

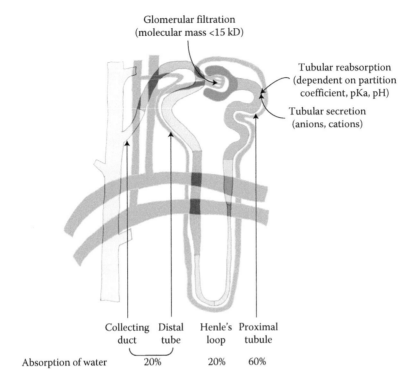

Glomerular filtration
(molecular mass <15 kD)

Tubular reabsorption
(dependent on partition
coefficient, pKa, pH)

Tubular secretion
(anions, cations)

	Collecting duct	Distal tube	Henle's loop	Proximal tubule
Absorption of water		20%	20%	60%

FIGURE 2.12
Schematic diagram of a nephron, the functional unit of the kidney. (Modified from Forth, W. et al., *Pharmakologie Und Toxikologie*, Spektrum Akademischer Verlag, Heidelberg, Germany, 1996; written by Fichtl et al., 1996.)

Albumin and hemoglobin are 60 kDa or more in molecular weight and cannot cross the glomerular capillary wall. However, compounds with a molecular weight of 15 kDa or less are filtered by glomerular capillary at the same rate as water in the absence of plasma protein binding. After filtration, these drugs are reabsorbed through the tubular pathway. Taking into account that renal elimination is ~1.5 L/day, a GFR of 120 mL/min (or 170 L/day) will culminate in the tubular reabsorption of ~99% of the drug (Figure 2.13b). Thus, ionic compounds and large molecules that cannot cross the renal epithelium will collect in the renal tube for secretion into the urine. In this case, the renal clearance is equivalent to the GFR.

Lipophilic substances, for their part, can be reabsorbed by the renal tube. Therefore, these substances are excreted into the urine at a lower rate than they are filtered, and the renal clearance becomes lower than the GFR (Figure 2.13c). Many endogenous substances are secreted through glomerular capillaries and the tubular lumen by passive transportation. In this case, the renal clearance is higher than the GFR (Figure 2.13d).

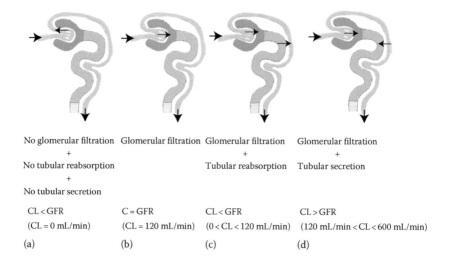

No glomerular filtration   Glomerular filtration   Glomerular filtration   Glomerular filtration
            +   +   +
No tubular reabsorption   Tubular reabsorption   Tubular secretion
            +
No tubular secretion

CL < GFR   C = GFR   CL < GFR   CL > GFR
(CL = 0 mL/min)   (CL = 120 mL/min)   (0 < CL < 120 mL/min)   (120 mL/min < CL < 600 mL/min)

(a)   (b)   (c)   (d)

FIGURE 2.13
Renal clearance as the result of the GFR, tubular reabsorption, and tubular secretion. Pathways of elimination. Foreign compounds are eliminated from the body by excretion and metabolism (biotransformation). The metabolites are ultimately removed from the body by excretion. (a) Clearance of >5 nm nanomedicines, (b) clearance of traditional drugs without tubular reabsorption/secretion, (c) clearance of traditional drugs with tubular reabsorption, and (d) clearance of traditional drugs with tubular secretion. (Modified from Forth, W. et al., *Pharmakologie Und Toxikologie*, Spektrum Akademischer Verlag, Heidelberg, Germany, 1996; written by Fichtl et al., 1996.)

An intact nanomedicine behaving the as same as a prodrug cannot take the same elimination path through the kidney as albumin or hemoglobin. Only released drugs or biodegraded fragments exhibit this attribute.

References

Fichtl, von B., G. Fulgraff, H. G. Neumann, P. Wollenberg, Forth sowie W., D. Henschler, and W. Rummel. 1996. Allegemeine pharmakologie und toxikologie. In *Allgemeine und spezielle Pharmakologie und Toxikologie: Für Studenten der Medizin, Veterinärmedizin, Pharmazie, Chemie, Biologie sowie für Ärzte, Tieräzte und Apotheker*, W. Forth, D. Henschler, W. Rummel, and K. Starke (editors), 6th edition. Heidelberg/Berlin/Oxford: Spektrum Akademischer Verlag, pp. 1–95.

Forth, W., D. Henschler, W. Rummel, and K. Starke. 1996. *Pharmakologie Und Toxikologie*. Heidelberg, Germany: Spektrum Akademischer Verlag.

Fujita, H. and T. Fujita. 1992. *Textbook of Histology: Part 2*, 3rd edn. Tokyo, Japan: Igaku-Shoin Ltd.

Hashizume, H., P. Baluk, S. Morikawa, J. W. McLean, G. Thurston, S. Roberge, R. K. Jain, and D. M. McDonald. 2000. Openings between defective endothelial cells explain tumor vessel leakiness. *American Journal of Pathology* 156 (4): 1363–1380. doi: 10.1016/S0002-9440(10)65006-7.

Hobbs, S. K., W. L. Monsky, F. Yuan, W. G. Roberts, L. Griffith, V. P. Torchilin, and R. K. Jain. 1998. Regulation of transport pathways in tumor vessels: Role of tumor type and microenvironment. *Proceedings of the National Academy of Sciences of the United States of America* 95 (8):4607–4612.

Igarashi, E. 2008. Factors affecting toxicity and efficacy of polymeric nanomedicines. *Toxicology and Applied Pharmacology* 229 (1):121–134. doi: 10.1016/j.taap.2008.02.007.

Klaassen, C. D. 1995. *Casarett and Doull's Toxicology: The Basic Science of Poisons*, 5th edn. New York: McGraw-Hill.

Maeda, H. and Y. Matsumura. 1989. Tumoritropic and lymphotropic principles of macromolecular drugs. *Critical Reviews in Therapeutic Drug Carrier Systems* 6 (3):193–210.

Marquardt, H., S. G. Schafer, R. O. McClellan, and F. Welsch. 1999. *Toxicology*. San Diego, CA: Academic Press.

Matsumura, Y. and H. Maeda. 1986. A new concept for macromolecular therapeutics in cancer-chemotherapy—Mechanism of tumoritropic accumulation of proteins and the antitumor agent smancs. *Cancer Research* 46 (12):6387–6392.

McNeil, S. E. 2009. Nanoparticle therapeutics: A personal perspective. *Wiley Interdisciplinary Reviews—Nanomedicine and Nanobiotechnology* 1 (3):264–271. doi: 10.1002/Wnan.006.

Morikawa, S., P. Baluk, T. Kaidoh, A. Haskell, R. K. Jain, and D. M. McDonald. 2002. Abnormalities in pericytes on blood vessels and endothelial sprouts in tumors. *The American Journal of Pathology* 160 (3):985–1000. doi: 10.1016/S0002-9440(10)64920-6.

Rhodin, J. A. and H. Fujita. 1989. Capillary growth in the mesentery of normal young rats. Intravital video and electron microscope analyses. *Journal of Submicroscopic Cytology and Pathology* 21 (1):1–34.

Uchino, H., Y. Matsumura, T. Negishi, F. Koizumi, T. Hayashi, T. Honda, N. Nishiyama, K. Kataoka, S. Naito, and T. Kakizoe. 2005. Cisplatin-incorporating polymeric micelles (NC-6004) can reduce nephrotoxicity and neurotoxicity of cisplatin in rats. *British Journal of Cancer* 93 (6):678–687. doi: 10.1038/sj.bjc.6602772.

Yamada, Y. 1994. *Modern Textbook of Histology, Third Edition*. Tokyo: Kanehara & Co., Ltd.

3

Respiratory and Olfactory Routes

Nanoparticles administered via the respiratory and olfactory routes mainly include three types of products: (1) nanocosmetics designed for the perfume or fragrance, (2) nanomedicines designed for medical therapy and diagnosis, and (3) nanomaterials designed for various consumer applications. This chapter will primarily focus on the administration and disposition of nanomedicines versus conventional drugs and chemical substances. Nanomedicines have recently been developed for the treatment of respiratory diseases, and the respiratory and olfactory routes provide promising alternatives to systemic delivery while largely bypassing hepatic metabolism. In addition, this chapter will discuss the occupational or environmental exposure of humans to nanomaterials via inhalation because an understanding of the fate of industrial nanomaterials enables an expectation of the fate of nanomedicines, especially those that fall into the nonbiodegradable category.

3.1 Anatomy and Physiology

3.1.1 Respiratory System

Anatomically, the respiratory system comprises the extrathoracic airway, the trachea, the lungs, and the pulmonary lobule, as defined by the International Commission on Radiological Protection (ICRP) in 1994. Functionally, the respiratory system is divided into (1) the conducting area, composed of the nasal cavity, nasopharynx, larynx, trachea, bronchi, bronchioles, and terminal bronchioles, and (2) the respiratory area, composed of the respiratory bronchioles, alveolar ducts, and alveoli (Mescher, 2010). The extrathoracic airway is in turn composed of the nasal cavity, nasopharynx, oropharynx, and larynx (Figure 3.1a), while the nasal cavity is separated into the right and left halves by the nasal septum. Each side of the nasal cavity contains its own olfactory and respiratory areas.

In humans, the respiratory area accounts for the majority (90% or more) of the nasal cavity. The total area of the human nasal cavity is 150–160 cm^2, while the total olfactory area is 10 cm^2 (Fujita and Fujita, 1992; Mescher, 2010). Thus, the olfactory area only occupies 3%–6% of the nasal cavity in humans. The nasopharynx is located in the middle of the pathway, from the nasal cavity to the esophagus, and receives air or liquid contents from the nasal cavity.

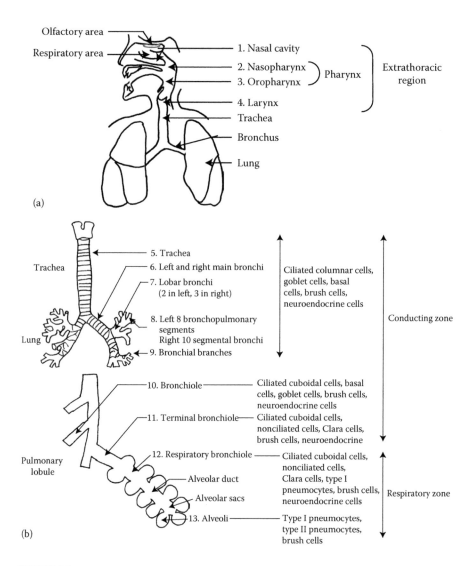

FIGURE 3.1
(a) The respiratory system. Anatomy of the human respiratory system showing the extra-thoracic airway. (b) The respiratory system. Anatomy of the trachea, lungs, and pulmonary lobule. The conducting zone composes 1–12, while the respiratory zone composes 12 and 13. (Modified from Yamada, Y., *Modern Textbook of Histology*, 3rd edn., Kanehara & Co., Ltd., Tokyo, Japan, 1994; Standring, S., *Gray's Anatomy*, 40th edn., Churchill Livingstone, Elsevier, London, 2008; Ross, M.H. and Pawlina, W., *Histology: A Text and Atlas*, 6th edn., Lippincott Williams & Wilkins, Baltimore, MD, 2011.)

The pathway near the oral cavity, the epithelium of the oral cavity to the larynx, receives the direct effect on air environment. For an example of dry environment, the inhalation of dry or cool air lowers the content of liquid on mucosa. The lowered liquid should affect the transport of substance and mucosal absorption. The decreased deposit of substances of the oral cavity to the larynx in air pathway affects the deposit into the alveoli. The decreased deposition of substance into the oral cavity to the larynx may elevate the deposit into the trachea to the alveoli.

The respiratory area in the nasal cavity comprises the superior, middle, and inferior nasal meatus and is lined with nasal mucosa. The nasal mucosa contains five types of cells: the ciliated columnar cells; mucous cells or goblet cells; brush cells; neuroendocrine cells or small granule cells, which are part of the diffuse neuroendocrine system; and basal cells (Standring, 2008; Ross and Pawlina, 2011). The nasal cavity and trachea both function as passageways for airflow entering the lungs (Figure 3.1b).

The epithelium extending from the primary bronchi to the bronchial branches contains ciliated columnar cells that act as a mucociliary escalator, as well as seromucous gland cells (also termed goblet cells), basal cells, brush cells, and neuroendocrine cells (also termed small granule cells). The latter are the respiratory representatives of a general class of enteroendocrine cells found in the gut and gut derivatives (Standring, 2008; Mescher, 2010). In humans, the number of generations from the trachea to the final bronchial branch is 23 when the trachea is assumed to be generation zero. There are fewer generations in the rat than in the human (Miller et al., 1993).

The bronchiolar epithelium consists of ciliated cuboidal cells, basal cells, seromucous gland cells, brush cells, and neuroendocrine cells (Standring, 2008; Mescher, 2010). Clara cells are located in the distal airways of the pulmonary lobule and increase along the length of the terminal bronchiole, while ciliated cells decrease along the length of the terminal bronchiole. The terminal bronchiole epithelium consists of ciliated cuboidal cells, nonciliated cells, Clara cells, brush cells, and neuroendocrine cells (Standring, 2008; Mescher, 2010). The epithelium of the respiratory bronchiole is composed of the same cell types as the terminal bronchiole epithelium, with the addition of type I pneumocytes (Standring, 2008). Lastly, the alveoli located in the respiratory bronchiole are formed of thin squamous type I alveolar cells, which compose 95% of the entire alveolar lining; cuboid type II alveolar cells, which compose ~5% of the lining; and a small number of brush cells occasionally found in the alveolar wall (Standring, 2008; Ross and Pawlina, 2011). In addition, alveolar macrophages function in the alveolar space to trap inhaled particulate materials from the pulmonary airways.

Following human exposure to nanoparticles via inhalation, the particle first contacts the nasal mucosa epithelium and then travels throughout the pulmonary lobule in the order shown in Figure 3.1, from the pharynx to the alveoli. Alveolar diameters in human and rat are approximately 200 and 100 μm (Miller et al., 1993; Sznitman, 2013).

The lamina propria is located underneath the mucous epithelium and is abundantly supplied with blood vessels. Lamina propria capillaries in the nasal mucosa are of the fenestrated type. Therefore, if nanoparticles pass the mucosal epithelial barrier, they travel through the microvasculature via fenestrated pores. In addition, lymphatic tissues (e.g., mucosa-associated lymphoid tissue [MALT]) are present in the lamina propria of the tracheal wall. MALT-like regions in the lung are termed bronchus-associated lymphoid tissue (BALT) and are the counterparts of gut-associated lymphoid tissue (GALT) in the GI tract. BALT is quantitatively limited in humans and primarily observed under pathological conditions, such as viral or bacterial infections (Levy et al., 2005). Therefore, nanoparticles are readily captured and transported across the epithelial barrier by M cells in the BALT during the induction of immune responses.

Brush cells or solitary chemosensory cells are epithelial cells distributed in (1) the nasal respiratory mucosa, the vomeronasal duct, and the auditory tube of the upper respiratory tract, (2) the olfactory mucosa, (3) the larynx, (4) the trachea and the bronchi of the lower airways, and (5) the alveolar region (only rat) (Krasteva and Kummer, 2012). Brush cells express molecular components of the canonical sweet, umami, and bitter taste signal transduction pathways (Behrens and Meyerhof, 2011). The function of the brush cells in relation to nanoproducts is discussed in Section 3.8.

The cilia of ciliated cells in the nasal and bronchial epithelia beat at reported rates of 12–16 times/s (Standring, 2008), 900–1200 strokes/min (Levy et al., 2005) or ~1000 strokes/min (Mygind and Dahl, 1998), and the rate is increased by mechanical stimulation as well as inflammatory stress (Standring, 2008). Ciliary beating in the nasal epithelium occurs in a downward motion to move mucus into the throat cavity, while the direction of ciliary beating in the bronchial epithelium is forward toward the throat cavity for clearance from the lungs. A study using normal human subjects showed that 80% of inhaled substances moved a distance of 3–25 mm/1 min outflow due to ciliary beating (Mygind and Dahl, 1998). The remaining 20% moved more slowly, with attenuated transport related to the properties of the inhaled airborne substance and lowered temperatures in the mucosa due to cold weather (Mygind and Dahl, 1998). Another study showed that epithelial ciliary-mediated movement of inhaled substances in the trachea and the bronchi was 5.5 and 2.4 mm/min, respectively, for normal human subjects (Foster et al., 1982).

The lungs receive the entire quantity of blood from cardiac output, or 5700 mL/min. This amount is nearly fivefold higher or more than that circulating to the hepatic portal, or 1125 mL/min, suggesting that drugs absorbed by the lungs are rapidly eliminated. The airway from the trachea to the alveoli continues to branch, while the diameter of the airway progressively decreases as branching increases, such that the surface area of the epithelium expands. The surface area of the alveoli is 100 m^2 or more, which is larger than the surface area of the small intestine. However, the total content of alveolar fluid (surfactant) is limited to only ~10 mL. The surfactant is secreted by type II alveolar cells and mainly consists of phospholipids (dipalmitoyl phosphatidylcholine and phosphatidylglycerol). Its low volume implies that inhaled, slowly soluble/insoluble or nonbiodegradable nanoparticles might accumulate in the lung for several hours or days in the absence of ciliated epithelial cells in the upper airway.

3.1.2 Olfactory System

The olfactory system in rodents and dogs includes the main olfactory bulb, the vomeronasal organ (VNO), and the accessory olfactory bulb, while the human olfactory system includes the main olfactory bulb and the poorly understood VNO. The function of the VNO is unclear in humans, and the existence of the accessory olfactory bulb is debatable.

The epithelium of the main olfactory bulb (Figure 3.2) is composed of four cell types, the cilia-bearing olfactory receptor (OR) cells, the solitary

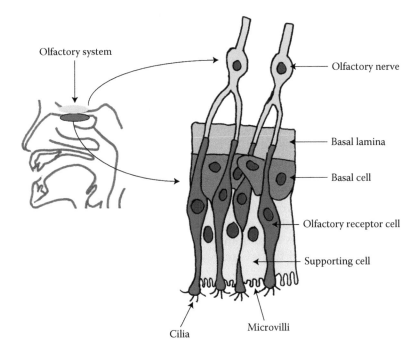

FIGURE 3.2
Macro- and microanatomy of the olfactory bulb. (Modified from Yamada, Y., *Modern Textbook of Histology*, 3rd edn., Kanehara & Co., Ltd., Tokyo, Japan, 1994; Ross, M.H. and Pawlina, W., *Histology: A Text and Atlas*, 5th edn., Lippincott Williams & Wilkins, Baltimore, MD, 2006; Standring, S., *Gray's Anatomy*, 40th edn., Churchill Livingstone, Elsevier, London, 2008; Adapted from data in Gloriam, D.E. et al., *BMC Genom.*, 8, 338, 2007, doi: 10.1186/1471-2164-8-338.)

chemosensory cells or brush cells (not shown in Figure 3.2), the microvilli-bearing supporting or sustentacular cells, and the basal cells (Ross and Pawlina, 2011). The main olfactory epithelium includes the olfactory gland, also called Bowman's gland.

When a deodorant substance first encounters the nasal cavity, it selectively binds to a 10–20 kDa odorant-binding protein secreted by supporting cells. The odorant-binding protein is delivered to OR cells in the main olfactory bulb via seven membrane-spanning G protein–coupled receptors (GPCRs), as well as to the VNO. The VNO is a bony-encased structure that lies bilaterally at the base of the septum and opens anteriorly into the nasal cavity or the mouth through the vomeronasal duct. The VNO is sensitive to common odorants and pheromones. After exposure of the organ to these compounds, the accessory olfactory bulb receives input from the ipsilateral VNO (Bigiani et al., 2005; Mucignat-Caretta, 2010; Yokosuka, 2012). The vomeronasal sensory epithelium contains solitary chemosensory cells that express taste-specific GPCRs (Bigiani et al., 2005; Mucignat-Caretta, 2010), whereas the accessory olfactory bulb epithelium contains microvilli-covered solitary chemosensory cells or brush cells, supporting cells, and basal cells (Fujita and Fujita, 1992; Mucignat-Caretta, 2010).

The percentage of the nasal epithelium that corresponds to the olfactory epithelium is higher in rats, mice, rabbits, and dogs than in monkeys or humans (Harkema et al., 2006). This value is ~50% in the rat versus only ~3% in the human (Harkema et al., 2006). Although the VNO is well developed in rodents and dogs, this organ is vestigial in human. Moreover, phylogenetic analyses of GPCRs in rat, mice, and human showed that the number of full-length OR and vomeronasal type I receptor (VIR) genes in rodents is higher than that in humans, and one-to-one orthologs are absent or rare (Gloriam et al., 2007). Furthermore, the gene expression level of G protein–coupled VIR receptors was threefold higher in mice than in humans (Table 3.1). No human/rat one-to-one VIR orthologs were detected, and the frequency of human/rat OR orthologs was only 8% (Gloriam et al., 2007). Taken together, these observations suggest that inhaled nanoparticles exert lower effects in humans than in mice or rats because nanoparticles must be absorbed to and endocytosed by OR and vomeronasal cells via GPCRs.

TABLE 3.1

Full-Length OR Genes in Mouse, Rat, and Human

	OR	VIR
Mouse	1081	145
Rat	1234	105
Human	846	53

3.2 Practical Application of Biodegradable Nanomedicines

Liposomes or polymer micelles are well-known platforms for the development of nanomedicines for therapeutic or diagnostic purposes. Many proteins found in the plasma have nanoscale dimensions, including insulin, cytokines, human growth hormone, albumin, and immunoglobulin (Ig) G (Figure 3.3). The engineering of nanomedicines loaded with nanosized proteins is thus an important focus of nanotechnology (Todoroff and Vanbever, 2011), in addition to the formulation of nanomedicines in engineered platforms composed of "generally recognized as safe" (GRAS) ingredients (Behl et al., 1998b).

Many platforms are engineered to be biodegradable due to safety concerns in the body. If biodegradability and other GRAS-associated issues are not considered during the experimental phase of nanoproduct development, the probability of their translation to the clinical phase is low. For example, metallic cadmium and lead are constituents of quantum dots (Oberdorster et al., 2005; Albrecht et al., 2006; Klaine et al., 2008; Sanvicens and Marco, 2008;

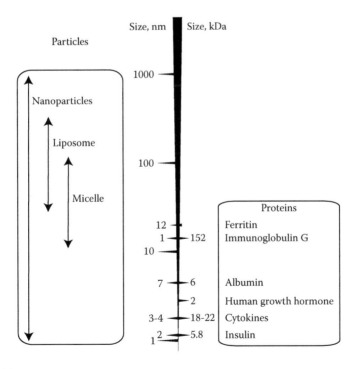

FIGURE 3.3
Particle size for nanomedicines and plasma proteins. (Modified from Todoroff, J. and Vanbever, R. *Curr. Opin. Colloid Interf. Sci.*, 16(3), 246, 2011, doi: 10.1016/j.cocis.2011.03.001.)

Aillon et al., 2009). However, cadmium features predominantly in acute and chronic toxicity, nephrotoxicity, pulmonary toxicity, carcinogenicity, hypertension, and adverse cardiovascular effects (Klaassen, 1995). Moreover, lead is known for its neurotoxicity and carcinogenicity, as well as its capacity to trigger peripheral neuropathy, hypertension, and adverse hematological, renal, and reproductive events (Klaassen, 1995). Therefore, quantum dots are inappropriate for therapeutic applications.

Biodegradable nanomedicines intended for oral administration or inhalation are currently designed with a nesting structure, similar to that of the Japanese playthings based on old folk art depictions of the seven gods of good fortune (Figure 3.4a). Recently, drugs for inhalation have been engineered as 1–5 μm delivery vehicle particles loaded with nanoparticles of

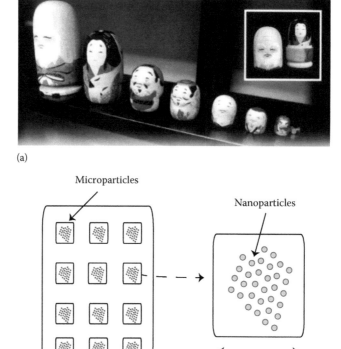

(a)

(b) Drug formulation

FIGURE 3.4
Nesting plaything and nesting structure of nanoproducts. (a) Japanese plaything based on old folk art depictions of Shichifukujin, or the seven gods of good fortune. The nesting plaything was initially manufactured in Japan more than 200 years ago and is the origin of the Russian nesting matryoshka doll. The figure was provided by Kazuyuki Tanaka in Hakone. (b) Nanomedicines for inhalation with a nesting structure, including a microscale vehicle enclosing nanoparticles.

≤1000 nm (Figure 3.4b). These drugs are generated by spray drying, emulsion, ionotropic gelation, or supercritical fluid extraction (Grenha et al., 2005; Chattopadhyay et al., 2007; Shi et al., 2007; Yamamoto et al., 2007; Bailey et al., 2008; Bailey and Berkland, 2009; Chattopadhyay et al., 2010; Morishita and Park, 2010; Narang and Mahato, 2010). Smaller nanoscale nanomedicines for inhalation without a nesting structure and lacking the microscale platform are not available in the present market.

3.3 Practical Application of Nonbiodegradable Nanomedicines

Nanodevices for medical implantation require material stability and are fabricated from nonbiodegradable nanomaterials. Nonbiodegradable nanomaterials are also used for engineering applications. Fullerenes, carbon nanotubes, and asbestos are all examples of nonbiodegradable nanomaterials that have been employed in the experimental phase. These nanomaterials all have the property of durability, but their exposure to humans is fraught with health risks, including an increased risk of malignant mesothelioma (Donaldson et al., 2010; Nagai and Toyokuni, 2010). Therefore, investigators are currently exploring the use of alternative, safer, durable nanomaterials.

3.4 Inhalation of Nanoproducts

Substances that are inhaled via both the nose and the mouth are eliminated by epithelial transport or macrophage phagocytosis before they arrive at the site of absorption. Therefore, the true dose of an inhaled substance or drug is not the inhaled dose, but is instead the bioavailable dose at the absorption site. The bioavailable dose, or the retained dose, is in turn the deposited dose minus the amount of cleared substance. Many drug losses occur between the initially metered dose and the retained dose. The metered dose is reduced by drug retention on instrumentation used to administer the medicine into the mouth or nose, nonspecific adsorption to cell surfaces, undefined losses, drug lost to exhalation, and drug rinsed away with water in the mouth. In addition, the large nanoparticle surface area enables particularly efficient adsorption onto the rough-surfaced walls of medical instruments or onto cell surfaces in the body, from the nose and mouth to the GI tract.

3.5 Transport and Liberation of Inhaled Products

Substances administered via inhalation include (1) gases and vapors and (2) nanoparticle aerosols. The physicochemical properties of nanoparticle aerosols are described by particle size, shape, solubility, hydrogen ion concentration, osmolarity, and density, where particle size is based on the concept of the aerodynamic diameter. These physicochemical properties critically affect nanoparticle deposition, retention, and mechanism of action in the lungs (Kuhn, 2001).

Nanomedicines and conventional medicines inhaled via the nose and mouth are transported in the airway and delivered to the site of absorption (Figure 3.5), where they are liberated from the deposited dose. The first step in pulmonary delivery is transport from the mouth or nose to the alveoli, which involves targeted delivery to the extrathoracic, tracheal, and alveolar compartments. Particle size is the most important factor affecting delivery and is expressed as the mass median aerodynamic diameter (MMAD). The MMAD is directly related to the absorption and distribution of the particle because particles consist of different types of shapes and sizes. The MMAD is defined as the diameter of a spherical particle with equivalent aerodynamic behavior.

Inhaled nanomedicines are liberated at the site of absorption via the release of nanoparticles and excipients from the microscale platform, followed by the release of active ingredients from the nanoparticles (Figure 3.5). The deposition of inhaled micro- and nanoparticles within a particular respiratory region (e.g., the extrathoracic region, trachea, bronchi, or alveoli) was previously estimated by the ICRP (1994) by using a dosimetric deposition model and a semiempirical approach. This model was applied to oral or nasal inhalation of unit density spheres at the reference-resting pattern for an adult Caucasian male sitting upright and awake (Cotes et al., 1979). The following five regions were considered particularly important for evaluating the regional disposition of inhaled particles: (1) the extrathoracic region in the anterior nose; (2) the posterior nasal passages of the pharynx, larynx, and mouth; (3) the upper tracheobronchial region, from the trachea to the bronchi, from which deposited substances are cleared by ciliary action; (4) the lower tracheobronchial region, from the proximal bronchiole to the terminal bronchiole; and (5) the alveolar interstitial region, from the respiratory bronchioles to the alveoli.

We can estimate the optimal size of a drug engineered for regional pulmonary delivery by the ICRP model, calculating for the polydispersity of aerosols to obtain reproducible results. Depending on the desired region, there are three peaks of optimal particle size for pulmonary delivery: ~1–5 nm, 5–500 nm, and 1 μm or more (Figures 3.6 and 3.7). However, the number of total deposits for microparticles of 1–10 μm is greater than that for nanoparticles of 100–999 nm. Thus, the weight of active ingredients for determining dose selection depends upon the engineering of the particle and is determined by the number of nanoparticles within a

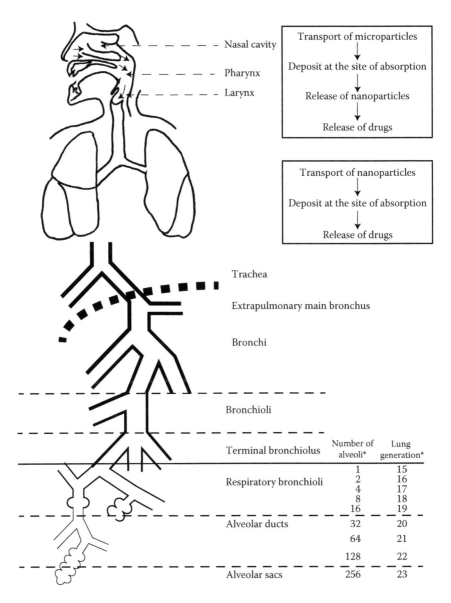

	Number of alveoli*	Lung generation*
Nasal cavity		
Pharynx		
Larynx		
Trachea		
Extrapulmonary main bronchus		
Bronchi		
Bronchioli		
Terminal bronchiolus		
	1	15
Respiratory bronchioli	2	16
	4	17
	8	18
	16	19
Alveolar ducts	32	20
	64	21
	128	22
Alveolar sacs	256	23

Transport of microparticles
↓
Deposit at the site of absorption
↓
Release of nanoparticles
↓
Release of drugs

Transport of nanoparticles
↓
Deposit at the site of absorption
↓
Release of drugs

FIGURE 3.5
Transport of particles from the nose or mouth to the trachea or alveoli. *Data are adapted from idealized dichotomous (symmetric) model of the pulmonary acinus for an average human adult scaled at functional residual capacity (Sznitman, 2013). (Modified from Bisgaard, H. et al., *Drug Delivery to the Lung, Lung Biology in Health and Disease*, Informa Healthcare USA, Inc., New York, 2007; Miller, F.J. et al., *Aerosol Sci. Technol.*, 18(3), 257, 1993, doi: 10.1080/02786829308959603; Sznitman, J., *J. Biomech.*, 46(2), 284, 2013, doi: 10.1016/j.jbiomech.2012.10.028.)

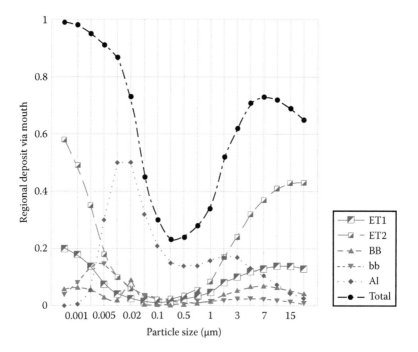

FIGURE 3.6

Regional deposition of inhaled particles within the lungs administered via the mouth in humans. The fractional deposition was calculated for a "mouth breather" assumed to take 60% of inspired air through the mouth. The mucosa of the tracheobronchial region is shown as the total of the sol layer and the lamina propria. ET1, the extrathoracic region in the anterior nose; ET2, the posterior nasal passages of the pharynx, larynx, and mouth; BB, the upper tracheobronchial region, from the trachea to the bronchi, from which deposited substances are cleared by ciliary action; bb, the lower tracheobronchial region, from the proximal bronchiole to the terminal bronchiole; and Al, the alveolar interstitial region, from the respiratory bronchioles to the alveoli. (Drawn from data in ICRP, ICRP publication 66, Human respiratory tract model for radiological protection, ICRP Annual Report, Vol. 24(1–3), 1994.)

nanoparticle-loaded microparticle, as well as the weight of the active ingredients in a single nanoparticle.

Targeted delivery to the tracheobronchial region is an ideal therapeutic modality for treating cystic fibrosis (Touw et al., 1995). The optimal particle size for targeted delivery to the tracheobronchial region depends on the particular tracheobronchial subregion (upper versus lower) and the route of administration. There are three major ICRP-predicted peaks for the most favorable tracheobronchial targeting: 1–10 nm via the nose or mouth, with two peaks of 10 nm or less via the mouth; 10–50 nm via the mouth; and ≥1 μm via the nose or mouth (Figures 3.6 and 3.7). Microscale particles inhaled via the mouth are mainly deposited in the upper tracheobronchial region, while nanoscale particles inhaled via the mouth are mainly deposited in the lower tracheobronchial region (Figure 3.6).

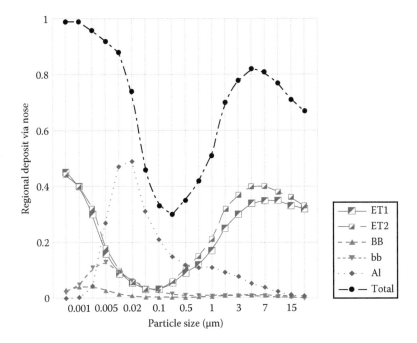

FIGURE 3.7

Regional deposition of inhaled particles within the lung administered via the nose in humans. The fractional deposition was calculated for a "nose breather" assumed to take 60% of inspired air through the nose. The mucosa of the tracheobronchial region is shown as the total of the sol layer and the lamina propria. ET1, the extrathoracic region in the anterior nose; ET2, the posterior nasal passages of the pharynx, larynx, and mouth; BB, the upper tracheobronchial region, from the trachea to the bronchi, from which deposited substances are cleared by ciliary action; bb, the lower tracheobronchial region, from the proximal bronchiole to the terminal bronchiole; and Al, the alveolar interstitial region, from the respiratory bronchioles to the alveoli. (Drawn from data in ICRP, ICRP publication 66, Human respiratory tract model for radiological protection, ICRP Annual Report, Vol. 24(1–3), 1994.)

The ICRP model predictions show a good correlation with the results of regional lung depositional studies of tobacco smoke and epidemiological studies of radon exposure. The occurrence of lung cancer is largely attributed to exposure to cigarette smoke (Pirozynski, 2006) and tobacco smoke particles of 100–300 nm (Robinson and Yu, 2001; Hofmann et al., 2009; McAughey et al., 2009). Moreover, the percentage of new patients presenting with small- and non-small-cell lung cancer is approximately 15% and 85% (Jemal et al., 2010). Meanwhile, small-cell lung cancer originates from neuroendocrine cells that are not located in the alveoli, but the 100–300 nm tobacco smoke nanoparticles distribute mainly to the alveolar region (Nazaroff et al., 1993; Hofmann et al., 2009). These results are consistent with the ICRP model and indicate that non-small-cell lung cancer is primarily attributed to the high disposition of tobacco nanoparticles in the alveoli. By contrast, the ICRP annual report demonstrated that small-cell lung cancer was the main type of lung cancer induced by radon (ICRP, 1994), suggesting the involvement of the tracheobronchial region.

The major deposition mechanisms of inhaled substances in the extrathoracic, tracheobronchial, and alveolar regions include impaction, sedimentation, interception, and diffusion (Figure 3.8a). Directional changes in airflow, airflow velocity, cross-sectional area, and residence time all affect particle deposition. Impaction, sedimentation, and interception are influenced by the aerodynamic properties of particles, while diffusion is affected by their Brownian motion. Particle impaction occurs when a particle moves away from the airway cavity and contacts the respiratory tract epithelium located in the airway wall. Impaction takes place at a sharp angle, at a branch point of the nasal cavity or the tracheobronchial region, or in contact with the ciliated cells of the epithelium, and is positively impacted by high particle velocity.

The lower tracheobronchial region branches more frequently than the upper tracheobronchial region and also has a larger number of shorter braches (Figure 3.5). Therefore, the lower tracheobronchial region undergoes marked directional changes in airflow, along with attenuated airflow velocity. Sedimentation of particles prevails under conditions of restricted airflow, as in the alveoli, where the rate of sedimentation is determined by the transport speed of the particle.

Particle interception plays a significant role for nanoparticles, but only for fiber particles having a high aspect ratio between the longest diameter and the shortest diameter. Interception takes place when the fiber particle accumulates at the edge of a tracheobronchial branch or along the wall of the nasal cavity and partially or completely obstructs the local airway. Longer fibers increase the probability of interception due to frustrated phagocytosis because macrophages cannot adequately extend themselves to enclose the particle (Donaldson et al., 2010).

Diffusion is the main process for deposition of particles of 500 nm or less and is common in the nasopharynx and alveoli. As particle size decreases, diffusion-mediated mobility increases, allowing particles to contact the surface of the airway. The deposition of particles of ≤100 nm occurs by Brownian motion and delivers particles to the alveolar region, while the deposition of particles of ≤10 nm is highly influenced by diffusion in the tracheobronchial region (Heyder et al., 1986). Airflow can change a fiber-shaped nanoparticle into one with a circular shape, depending on the strength or elasticity of the fiber. In this regard, carbon nanotubes are known to exhibit flexible bending properties.

The lung displays species-specific anatomical differences in the branching pattern and alveolar diameter (Figure 3.8b). The anatomical differences may change the effect of inhaled particle by the aerodynamic parameters as impaction and interception.

Size-dependent deposition is characterized by thermodynamic, intermediate, and aerodynamic domains (Figure 3.9) (ICRP, 1994). Deposition of

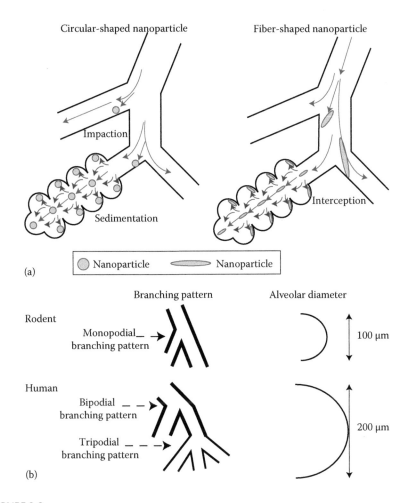

FIGURE 3.8
Aerodynamic mechanisms of the deposition of inhaled particles in the respiratory tract and two factors affect into the mechanisms in comparisons between rodents and humans. (a) Aerodynamic mechanisms of two nanoparticles, circular-shaped nanoparticle and fiber-shaped nanoparticle, and (b) two factors of branching pattern and alveolar diameter affect into the mechanisms in rodent and human. (a: Modified from aerodynamic mechanisms by Marquardt, H. et al., *Toxicology*, Academic Press, San Diego, CA, 1999 and airflow analysis in the alveolar region by Li, Z. and Kleinstreuer, C., *Med. Biol. Eng. Comput.*, 49(4), 441, 2011, doi: 10.1007/s11517-011-0743-1; b: Modified from data in Miller, F.J. et al., *Aerosol Sci. Technol.*, 18(3), 257, 1993, doi: 10.1080/02786829308959603.)

nanoparticles of ≤100 nm is affected by the thermodynamic domain, while deposition of microparticles of 1 μm or more is affected by the aerodynamic domain. Based on this principle, a nanoparticle diameter of ≤100 nm is considered a thermodynamic diameter, while a microparticle diameter of ≥1 μm is considered an aerodynamic diameter.

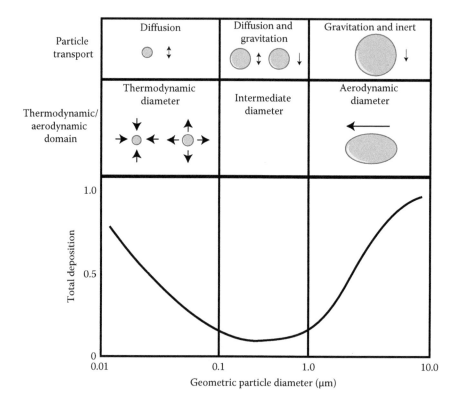

FIGURE 3.9
Deposition of inhaled particles showing the aerodynamic and thermodynamic domains of
inhaled products. (Modified from Marquardt, H. et al., *Toxicology*, Academic Press, San Diego,
CA, 1999.)

Deposition of particles occurs within the airway and on the epithelial
surface by diffusional transport within the thermodynamic domain. Total
deposition diminishes as a function of gravity as particle size increases to
≥10 μm. Changes in total deposition are influenced by increases or decreases
in breathing rate, but not by the density of the particles or by the aerosol flow
rate, because the effects of gravity on density are low.

Deposition of 100–1000 nm particles occurs within the airway and on the
epithelial surface by both diffusional and gravitational particle transports in
the intermediate domain. When diffusional transport is equivalent to gravitational transport, total deposition is at its minimum. However, deposition of
1–10 μm particles occurs within the airway and on the epithelial surface by
gravitational and inertial transports in the aerodynamic domain. Total deposition increases together with increases in particle size and density, aerosol
flow rate, and breathing rate.

3.6 Absorption via the Alveoli

Absorption of particles in the alveoli occurs earlier than mucosal uptake in the nasal and oral cavities, intestine, and rectum. Alveolar absorption occurs most readily for low-molecular-weight (LMW) (100–1000 Da) lipophilic substances via diffusion after permeation of the lipid membrane, while high-molecular-weight (HMW) substances depend on transport factors for membrane penetration. For example, peak levels in the blood of soluble LMW proteins were observed at 10–30 min after injection, whereas peak levels of HMW (>1 kDa) proteins were observed after a period of days (Lombry et al., 2004). The mechanisms of penetration into the alveolar epithelial cell include paracellular transport via epithelial intercellular gaps and transcellular transport via epithelial pores for proteins of >1 and <40 kDa (Patton, 1996), as opposed to transcellular transport via endocytosis for proteins of 40 kDa or more (Matsukawa et al., 1997).

Charge also affects the transport of HMW proteins (Bitonti and Dumont, 2006; Yacobi et al., 2008, 2010). Anionic HMW proteins are generally not imported into cells by endocytosis, but cationic proteins are.

3.7 Distribution and Translocation

3.7.1 Biodegradable Nanoparticles

Transport of HMW substances occurs via columnar epithelial cells in both tracheal and thin alveolar epithelial cells (Bur et al., 2006; Grainger et al., 2009). The epithelial lining of the upper respiratory tract sets up a tight barrier to the transport of these substances into the circulatory system. Transport or absorption of HMW substances occurs at a relatively high rate in the lower alveolar airway compared with the upper tracheal airway due to the large surface area of the alveoli. Rapid absorption of biodegradable HMW substances by the alveoli mitigates long-term exposure and degradation in the trachea, thereby augmenting systemic distribution of the active drug. Heparin is an example of a stable drug that is absorbed by both the trachea and the alveoli.

Protein, drug, or nanoparticle delivery via the alveolar epithelium occurs by paracellular or transcellular transport, depending on the size of the compound or particle. Alveolar epithelial cells do not afford size-dependent transport by diffusion through cellular gaps. Although numerous substances, such as glucagon-like peptide 1, albumin, transferrin, and IgG, are conveyed into these cells by transcellular transport, others (e.g., parathyroid, insulin, and growth hormones) are not. However, insulin can be imported into alveolar epithelial cells by transcytosis via the insulin receptor. The cellular uptake of insulin is terminated once a clinically effective concentration is attained, implying that insulin is primarily transported across the epithelium via the paracellular pathway. Lastly, substances with no specific receptor in alveolar epithelial cells are imported via nonspecific pinocytosis or paracellular diffusion. The 22 kDa growth hormone is transported by diffusion, whereas the 40 kDa horseradish peroxidase is imported by nonspecific pinocytosis.

Translocation from the lung to the blood is the least-utilized pathway affecting the fate of administered drugs and nanoparticles (Takenaka et al., 2006; Moller et al., 2008; Furuyama et al., 2009). In a recent study, normal healthy subjects were administered technetium (^{99m}Tc)-radiolabeled, 100 nm carbon particles via inhalation (Moller et al., 2008). Approximately 25% of the deposited particles were cleared within 1 day from the airways of healthy nonsmokers by mucociliary action, while ~75% of the particles were persistent. No nanoparticle-derived radioactivity was detected in the liver, suggesting that translocation to the blood did not occur. Analysis of the activity profiles obtained from gamma camera images of the abdomen also showed no significant difference from the background profile.

In another study, gold- or fluorescein-labeled polystyrene nanoparticles were administered to mice via intrathecal administration (50 µg/mouse for 20 and 200 nm gold nanoparticles and 200 µg/mouse for 200 nm, fluorescein-labeled nanoparticles) (Furuyama et al., 2009). The nanoparticle distributions were then studied in the trachea and alveoli over time. Gold was found in the alveoli at 2 h after administration for both nanoparticle sizes. Tissue distribution analysis at the same time revealed 75.2% of the gold label in the lungs, 0.01% in the liver, 0.003% in the spleen, and 0.002% in the kidney for the 20 nm particles, compared with 89.6% in the lungs, 0.025% in the liver, 0.0008% in the spleen, and 0.002% in the kidney for the 200 nm particles. No significant differences were found between the tissue distributions for 20 versus 200 nm particles. Colloidal gold nanoparticles were detected in both type I and type II alveolar epithelial cells, as well as in alveolar macrophages, suggesting that the nanoparticles crossed the alveolar epithelium via the transcellular pathway and were transported into the extrapulmonary blood vessel via macrophage-mediated endocytosis.

An additional tissue distribution analysis investigated the fate of 16 nm gold nanoparticles (88 µg/m^3) administered to rats by inhalation for 6 h (Takenaka et al., 2006). The primary gold particles collected from the aerosol on day 7 were 5–8 nm in diameter and spherical, and the agglomerates showed an average length of 27.7 nm. An inductively coupled plasma mass spectrometry (ICP-MS) analysis of gold concentration and content in lung and blood at 7 days after administration revealed a high overall concentration of gold nanoparticles (947 µg/kg), with 78% of the particles in the lungs and none in the blood. Furthermore, morphological analysis at 7 days after inhalation revealed gold nanoparticles mainly in the cytoplasmic vesicles of type I alveolar epithelial cells and macrophages and occasionally in endothelial cells and the septal interstitium.

Choi and colleagues investigated the translocation of inhaled nanoparticles from the alveolar space to the systemic circulation by using inorganic/organic hybrid nanoparticles and organic nanoparticles of a variety of shapes, sizes, charges, and physicochemical properties (Choi et al., 2010). Organic nanoparticles with a dynamic diameter of ≤34 nm and a noncationic surface after 1 h in serum translocated from the alveolar epithelium to the septal interstitium and then into the blood via the lymph nodes. Inorganic/organic nanoparticles with a dynamic diameter of ≤6 nm after 1 h in serum rapidly translocated from the alveolar epithelium to the blood via the lymph nodes, followed by elimination via the kidney. Lastly, an in vitro study of 6 nm dextran nanoparticles showed that 6 nm was the upper size for alveolar epithelial paracellular transport via diffusion (Matsukawa et al., 1997).

3.7.2 Nonbiodegradable Nanoparticles and Piercing

Durability is closely related to the biopersistence of nonbiodegradable nanoparticles. The relationship between durability and pathogenicity of four types of carbon nanotubes was compared with that of asbestos fibers (Osmond-McLeod et al., 2011). The various substances were injected into the peritoneal cavity of mice. Nanotubes with a lower mass or a shorter length were associated with a loss of pathogenicity, but rope-like aggregates of intact nanotubes with sufficient length and a high aspect ratio induced an asbestos-like response in mice.

Nagai et al. (2011) explored the link between fiber products as a function of size, cell membrane penetration, and the induction of malignant mesothelioma. Cell membrane penetration by asbestos fibers and different types of multiwalled carbon nanotubes (MWCNTs) (5 μg/cm^2 of each) was examined by assessing the flow cytometric side scatter value in cultured MeT5A and HPMC mesothelial cells, macrophages, and epithelial cells. Both mesothelial cell lines and the macrophages showed a marked increase in penetration after incubation for 3 h with asbestos. The mesothelial cells showed no increase in penetration with any of the MWCNTs, while the macrophages showed augmented penetration with all of the MWCNTs except for the tangled type. The penetration of asbestos into the cell membrane was associated with asbestos-induced cell death. The detrimental impact of the tangled-type and 143 nm MWCNTs was limited, but the 50 nm MWCNTs were highly cytotoxic. These findings suggest that the cytotoxicity of MWCNTs is inversely correlated with diameter.

The penetration mechanism of the 50 nm MWCNTs was classified as a "piercing" event, or an energy-independent penetration of the cell membrane, rather than an active internalization, or an energy-dependent import of foreign materials into the cell (Nagai and Toyokuni, 2012). This classification was based on the absence of a vesicular membrane structure associated with the pierced membrane. The 50 nm MWCNTs penetrated the nuclear membrane as well as the plasma membrane, whereas the 116 and 143 nm MWCNTs did not, demonstrating the diameter-dependent nature of membrane piercing by MWCNTs.

Injection of mice with 50 nm MWCNTs (1 mg) induced malignant mesothelioma with a higher frequency and earlier progression than injection with the same amount of 143 nm MWCNTs or 10 mg of tangled MWCNTs. Thus, the piercing event by MWCNTs is limited to nanotubes of 100 nm or less, indicating that the phenomenon lies within the normal range of pinocytosis. The phenomenon is not coincident with the range of the clathrin-mediated endocytosis, since the clathrin-mediated endocytosis is a receptor-mediated process and the cargo is 150 nm in size and circular in shape. It suggests lacking no receptor for the receptor-mediated endocytosis of MWCNT. The size and shape of cargo in almost pinocytosis except for clathrin- and caveolin-independent endocytosis, referred to as GEEC pathway, is 100 nm or less and circular or ellipse, while the size and shape of cargo in the GEEC pathway is 50 nm in shortest diameter (width) and 200–600 nm in longest diameter (depth) and tubular in pinocytosis (Figure 4.11). In addition, asbestos induced membrane-piercing events occur in the size of both 100 nm or less and 100 nm or more. The carbon nanotube of 100 nm or more in shortest diameter cannot be easily endocytosed over the size of cargo in shortest diameter in pinocytosis. Therefore, the membrane-piercing event of carbon nanotube is consistent with the shape and size of cargo in pinocytosis, whereas that of asbestos is not consistent with the shape and size of cargo in pinocytosis.

Broaddus et al. (2011) recently reviewed the mechanism of fiber translocation into mesothelioma cells in rodents. The route of translocation from the alveoli to the mesothelial cell is unknown because analysis of the pleural space is limited by the narrowness of the space and the difficulty in sampling without inducing inflammation or injury. Asbestos fibers are postulated to migrate to the lung interstitium and the visceral pleura by a paracellular pathway or by direct penetration across the injured alveolar epithelial cell. The fibers may be transported to the pleural space via the lymphatic system and the bloodstream, likely via passive translocation to the pleural space, followed by transfer to the partial pleura, intercostal microvessels, or lymphatic vessels. However, the ability to extrapolate from rodent studies to humans is limited because of differences between the structure of the visceral pleura in rodents and humans.

The visceral pleura in rodents has a thin wall consisting of a mesothelial layer and a basement membrane lying directly over the alveoli, while the visceral pleura in humans has a thick wall consisting of submesothelial connective tissue space, nerves, and systemic blood vessels. These discrepancies could potentially influence fiber movement into the pleural space. Although the pathological responses between rodents and humans are similar with respect to pulmonary fibrosis, the responses with respect to pleural fibrosis are not.

3.8 Disposition of Nanomedicines from the Nasal Cavity to Systemic Tissues

Nasal mucosal drug delivery via inhalation through the nose is utilized for therapy against local disease because absorption via the nasal route avoids the intestinal and hepatic pathways. This route of administration also provides an alternative to intravenous injection for the systemic delivery of drugs via blood capillaries. Nasal mucosal drug delivery is very simple and well tolerated by patients. The route sometimes yields bioavailability as low as 10%, but the nonintravenous nasal delivery system is still beneficial from the viewpoint of health-related quality of life and convenience.

The total volume of nasal secretions at any given time is 15 µL under normal physiological conditions. The available volume for nasal delivery is limited to 25–200 µL (Behl et al., 1998a). The factors affecting membrane permeability following nasal inhalation are the same as those in other drug administration routes and encompass biological factors such as anatomical structure; biochemical, physiological, pathological, and environmental factors; and pharmaceutical, physicochemical, and engineering factors, including particle size (Behl et al., 1998a,b; Donovan and Huang, 1998; Ahmed et al., 2000).

The nasal cavity comprises five anatomical regions: (1) the nasal vestibule, (2) the atrium, (3) the respiratory area, (4) the olfactory area, and (5) the nasopharynx (Figure 3.10). The permeability of each region influences drug bioavailability (Ugwoke et al., 2001; Arora et al., 2002). The nasal vestibule is the least permeable of the five regions due to the presence of keratinized cells. Likewise, the atrium is not easily permeable due to its small surface area and the presence of keratinocytes in the anterior portion. On the other hand, the respiratory area is the most permeable because of its large surface area and high density of blood capillaries. The olfactory area, composed of the main olfactory bulb, the accessory olfactory bulb, and the VNO, is particularly permeable because of its close proximity to the cerebrospinal fluid and its abundant expression of GPCRs. The nasopharynx is responsible for excretions from the nasal cavity and is also highly permeable.

Permeability is dependent on the cell types that constitute the nasal epithelium, in addition to cell density and number (Figure 3.10). The presence of cilia- and microvilli-bearing cells in the epithelium increases its surface area and permeability to drugs; secretion of gel-forming mucins by goblet cells also affects epithelial permeability.

Drug disposition after inhalation of nanomedicines via the nasal cavity has been investigated by using nonbiodegradable micro-/nanomaterials and biodegradable micro-/nanoparticles. Translocation is one of the most remarkable characteristics in disposition. Translocation after inhalation of micro-/nanoparticles in rats was shown in several studies of translocation to neurons via the olfactory bulb. For example, a study using microparticles

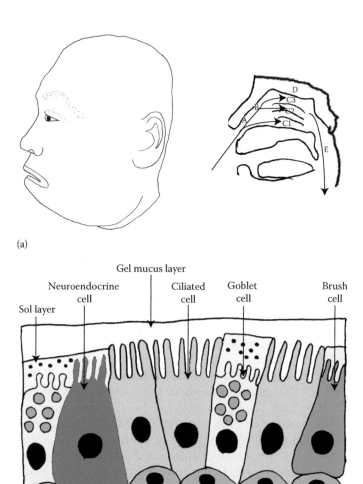

(a)

(b)

FIGURE 3.10
Nasal cavity and nasal mucosa: (a) Nasal cavity showing the (A) nasal vestibule; (B) atrium; (C) respiratory airway; (C1) superior meatus; (C2) middle meatus; (C3) inferior meatus; (D) olfactory bulb, including the accessory olfactory bulb and the VNO; and (E) nasopharynx. (b) Nasal mucosa. (Modified from Ugwoke, M.I. et al., *J. Pharm. Pharmacol.*, 53(1), 3, 2001.)

and nanoparticles showed a high distribution of biodegradable versus non-biodegradable microparticles or nanoparticles in the olfactory bulb, with a high probability of translocation of manganese dioxide microparticles to neurons (Fechter et al., 2002). These observations suggest that biodegradability is a factor in neuronal translocation.

Another inhalation study of 1.3 μm manganese dioxide microparticles in rats (Fechter et al., 2002) showed a significant increase in manganese levels (0.75 mg/g) in the cerebral cortex and a similar tendency in the olfactory bulb. The disposition in the lungs and translocation into neurons showed a particle size–dependent relationship, as follows. Rats were exposed to 1.3 and 18 μm manganese dioxide microparticles or a negative control for 6 h/day, 5 days/week, for 3 weeks by using a nose-only inhalation apparatus. The aerosol mass concentration of manganese dioxide was 4.75 mg/m^3 (or 3 mg/m^3 manganese) for both types of particles. Both the 1.3 and 18 μm particles showed increased distribution to the lungs compared with the control, but the 1.3 μm particle exhibited higher distribution than the 18 μm particle. When the manganese concentrations were analyzed according to brain region (i.e., olfactory bulb, cortex, cerebellum, brainstem, diencephalon, and basal ganglia), the highest manganese concentration attributed to the 1.3 μm particle (~2.5 mg/g) was found in the olfactory bulb. Nevertheless, this result was not significant compared with the concentration of the control (1.5 mg/g).

Increased manganese dioxide levels in the olfactory bulb were found in additional inhalation studies that employed 30 nm manganese dioxide nanoparticles and 3 nm radiolabeled carbon nanoparticles (Oberdorster et al., 2004; Elder et al., 2006). In the first study (Elder et al., 2006), manganese levels increased even in the deep regions of the central nervous system after nanoparticle administration, including the frontal cortex, corpus striatum, and cerebellum. Hence, the olfactory neuronal pathway efficiently translocated inhaled manganese dioxide nanoproducts to the central nervous system via the olfactory bulb. Nonetheless, the nanoparticles revealed the potential to trigger inflammatory changes during their travels along this pathway.

In the second study (Oberdorster et al., 2004), rats were exposed for 6 h to radiolabeled carbon nanoparticles at a concentration of 160 μg/m^3. The concentration of distributed isotope was determined by isotope radio mass spectroscopy in the lungs, olfactory bulb, cerebrum, and cerebellum. The concentration in the lungs decreased from 1.39 mg/g on day 1 to 0.59 μg/g on day 7. The corresponding concentrations in the olfactory bulb, cerebrum, and cerebellum were 4.3, 0.23, and 0.21 μg/g, respectively, suggesting the probability of translocation to the brain via the olfactory bulb. However, extrapolation from the rat to humans should be underestimated because administration and transport via the nasal olfactory bulb is highly different in the two species from a physiological and anatomical viewpoint. Namely, nanoproduct administration occurs via forced inhalation in rats and voluntary administration in humans. Also, the percentage of the nasal mucosa taken up by the olfactory region is 50% in the rat versus 5% in humans, and the weight of the olfactory bulb is 177-fold larger in the rat than in humans. Furthermore, the relative weight of the olfactory bulb per unit body weight is 85 ng/0.2 kg in the rat versus 168 ng/70 kg in humans (Oberdorster et al., 2005).

3.9 Clearance of Nanoparticles

Particles are cleared from the airway by four predominant mechanisms: (1) direct clearance via ciliated epithelial cells, (2) indirect clearance by ciliated epithelial cells, (3) clearance and degradation by alveolar macrophages, and (4) clearance by alveolar macrophages, followed by transport to the lung interstitium and the lymph vessels and nodes.

Deposited substances in the nasopharynx are translocated above the surface of cilial mucosa by sneezing or blowing. Alternatively, they are rapidly cleared by engulfment or salivary flow by reverse movement via the mucosal ciliary epithelium to the oral cavity (Figure 3.11). Alveolar macrophages represent 93% of the cells in normal adult lung lavage fluid; lymphocytes constitute ~7% of the remaining cells; and neutrophils, eosinophils, and basophilic leukocytes account for 1% or fewer (Hunninghake et al., 1979). Alveolar macrophages interact not only with bacteria or particles but also with polymers. Large particles are readily cleared from the lungs by phagocytosis.

Intratracheal instillation of polystyrene microsphere particles following inhalation showed early and late biphasic clearance (Snipes and Clem, 1981; Moller et al., 2004), as well as size-dependent translocation to the lymph nodes (Snipes and Clem, 1981). In early clearance, the half-lives of 50% for 3 μm particles and 24% for 9 μm particles were 12 and 17 days, respectively, whereas the half-life of 14% for 15 μm particles was 2.3 days. In late clearance, the half-lives of 50% for 3 μm particles and 76% for 9 μm particles were 69 and 580 days, respectively, while the half-life of 86% for 15 μm particles was not measurable during the 106 days of investigation. The 3 μm particles, but not the 9 and 15 μm particles, were translocated to the lung-associated lymph nodes.

Deposits in the tracheobronchial region are usually quickly cleared in a manner of hours (Figure 3.11). Translocation to the bronchioles and upper bronchi via the mucosal ciliary escalator is facilitated by substances secreted from macrophages. Clearance mechanisms for particle removal are likewise effective for the removal of bacteria and viruses, protecting the airway from infectious factors.

Clearance by macrophages is the main mechanism of elimination in alveolar regions having no ciliated cells (Figure 3.11), but macrophage-mediated clearance is slower than ciliary clearance and provokes accumulation of inhaled particles. A study of healthy, nonsmoking subjects analyzed particle clearance after inhalation of spherical, monodisperse ferromagnetic iron oxide particles with a 1.9 μm geometric diameter and a 4.2 μm aerodynamic diameter (Moller et al., 2004). Approximately 50% of the deposited particles were removed by fast clearance, with a half-life of 3 h after exposure, while the remaining 50% were removed by slow clearance, with a half-life of 109 days. The biphasic fast and slow clearance rates were indicative of mucociliary- and macrophage-mediated clearance, respectively. Particles

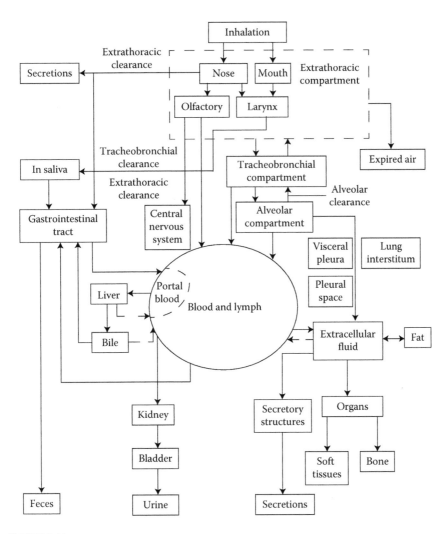

FIGURE 3.11
Clearance of inhaled nanoproducts. (Modified from Klaassen, C.D., *Casarett and Doull's Toxicology: The Basic Science of Poisons*, 5th edn., McGraw-Hill, New York, 1995.)

were then translocated to the interstitium and cleared through the lymph nodes located in the lungs.

3.9.1 Clearance by Mucosal Ciliated Cells

Extrathoracic or tracheobronchial mucosal ciliated cells can eliminate nano-products and nanomedicines that do not cross the mucosal barrier, as well as nanomedicines bound to mucosal compounds or suspended in mucosal liq-uid. However, mucosal ciliated cells do not assist in nanomedicine transfer across the mucosal barrier. This barrier is an epithelial mesh fiber with pores of 20–800 nm. A large particle that exceeds the size of the pores disappears into the gel layer and adheres to mucosal constituents by electrostatic interactions with carboxyl or sulfate groups, hydrophobic interactions, interactions with polymeric chains, or hydrogen bonding. IgG, with a diameter of 11 nm, diffuses as rapidly in mucosal liquid as in saline, while IgM demonstrates low affinity with mucosal constituents and delayed transport in mucosal liquid.

Nanoparticle deposit and clearance were studied in dogs following intratracheal administration of radiolabeled human blood plasma and 220 nm insoluble colloidal particles (Lay et al., 2003). Albumin was eliminated at a slower speed than the colloidal particles and subsequently accumulated in the mucosa. This observation signifies that albumin and other soluble substances with low membrane permeance attach to the surface of the airway epithelium more efficiently than insoluble substances. Therefore, because transport by the gel layer is no more effective than transport by the sol layer, albumin diffuses into the gel layer rather than into the sol layer and is cleared by mucosal ciliated cells.

Nanoparticles coated with polyethylene glycol (PEG) can rapidly permeate the mucosa (Lai et al., 2007). Lai and colleagues showed that PEG (2 kDa)-coated polystyrene nanoparticles were transported via diffusion in endocervicovaginal mucosa in a size-dependent fashion. The percentages of diffusive 100, 200, and 500 nm PEG-coated particles were ~8%, 62%, and 70%, respectively, whereas those of uncoated 100, 200, and 500 nm particles were ~1%, 10%, and 10%, respectively, signifying a delay in diffusion for uncoated nanoparticles. Therefore, proper coating of large nanoparticles expedites their penetration of physiological human mucus. Uncoated nanoparticles were captured in the mucosa because the hydrophobic polystyrene utilized for their manufacture enabled interactions with hydrophobic elements in the mucosa. However, PEG is water soluble and electrostatically neutral, thereby minimizing the interaction of PEG-coated nanoparticles with mucosa.

Particles coated with 5 kDa PEG are also highly mucosa permeant, but particles coated with 10 kDa PEG are not (Wang et al., 2008). This result suggests that 5–10 kDa is the transitional size for PEG in terms of membrane permeance. Indeed, 10 kDa PEG strongly adheres to the mucosa, probably due to associations between mucosal fibers and PEG, as well as to the formation of oxygen and hydrogen bonds between PEG and glucose in the mucosal fluid.

3.9.2 Clearance by Alveolar Macrophages

Although macrophages located at the surface of the alveoli effectively eliminate proteins from the airway space, the same is not true for nanoparticles. Thus, alveolar macrophages apparently do not recognize nanoparticles in the same way that they recognize proteins. While nanoparticles adhere to the epithelial surface, proteins are more likely to adsorb to the macrophage membrane. The critical significance of the ability of macrophages to recognize proteins is exemplified by pulmonary alveolar proteinosis, a rare disease in humans caused by the accumulation of lipoproteins in the alveoli.

Moller and colleagues showed that insoluble, nonbiodegradable carbon nanoparticles accumulated in the lungs over a period of several weeks and were then eliminated by the larynx/esophagus/intestine via the mucosal escalator. Human subjects were administered 100 nm carbon nanoparticles radiolabeled by ^{99m}Te via inhalation. Particle deposit, retention, and clearance were then assessed over time (Moller et al., 2008). Particle retention after 48 h was 95% in the alveoli and 70% in the trachea.

Retention of nanoparticles after inhalation is decreased by clearance via mucosal ciliated cells. Human subjects were administered 20 nm nanoparticles radiolabeled by iridium (^{192}Ir) via inhalation (Semmler-Behnke et al., 2007). Fragmentation of the nanoparticles decreased by 20%, while 80% of the nanoparticles were retained over a period of several months. The particles were subsequently eliminated in the feces or cleared from the lungs.

Phagocytosis by alveolar macrophages is the main mechanism of clearance of microparticles from the lungs. However, Geiser et al. (2008) demonstrated that phagocytosis cannot effectively clear nanoparticles of 200 nm or less. In this study, macrophage-mediated clearance was evaluated after the inhalation of 20 nm titanium dioxide nanoparticles. Engulfment of these particles was only ~0.1% within 24 h. By contrast, the particles were imported by macrophages after their incorporation into 3–6 μm vesicles containing surfactant and other substances. Agglomerated particles were not found in association with the macrophage vesicular membrane. These findings indicate that although macrophages preferentially uptake microparticles versus nanoparticles, nanoparticles can be imported together with substances subject to phagocytosis.

3.10 Nanoscale World According to the Respiratory and Olfactory Routes

The sizes of various inhaled nanoparticles, anatomical features, and permeability barriers in the respiratory and olfactory systems are provided by the descriptions in this chapter and illustrated in Figure 3.12. ICRP models of nanoparticle administration via the nose show that nanoparticles of ~5 nm or less accumulate in the nasopharynx, while nanoparticles of 7–700 nm (peak, 100–300 nm) accumulate in the alveoli, and microparticles of 1–10 μm accumulate in the nasopharynx at high levels and the tracheobronchial region at low levels. Extrapolation from the ICRP models and consideration of the anatomical and epidemiological data of lung cancer incidence allow an accurate expectation of the particle size/accumulation relationship for inhaled

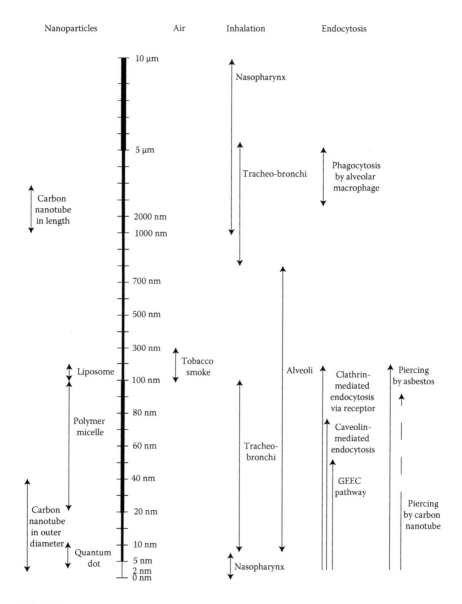

FIGURE 3.12
Size of nanoproducts and tobacco smoke and their disposition via the respiratory and olfactory routes.

nanoproducts. For example, the size of tobacco smoke nanoparticles is 100–300 nm, which matches the peak size of accumulation in the alveoli predicted by the ICRP model. On the other hand, the small size of carbon nanotubes (≤100 nm) allows them to be endocytosed by a clathrin-mediated or clathrin- and caveolin-independent mechanism. Liposomes are 100–200 nm in size, indicating that they are also likely to accumulate in the alveoli, while polymer micelles are 20–100 nm and are predicted to accumulate in the tracheobronchial region. Finally, microparticles are more efficiently cleared than nanoparticles via alveolar macrophage–driven phagocytosis.

References

Ahmed, S., A. P. Sileno, J. C. deMeireles, R. Dua, H. K. Pimplaskar, W. J. Xia, J. Marinaro, E. Langenback, F. J. Matos, L. Putcha, V. D. Romeo, and C. R. Behl. 2000. Effects of pH and dose on nasal absorption of scopolamine hydrobromide in human subjects. *Pharmaceutical Research* 17 (8):974–977.

Aillon, K. L., Y. M. Xie, N. El-Gendy, C. J. Berkland, and M. L. Forrest. 2009. Effects of nanomaterial physicochemical properties on in vivo toxicity. *Advanced Drug Delivery Reviews* 61 (6):457–466. doi: 10.1016/j.addr.2009.03.010.

Albrecht, M. A., C. W. Evans, and C. L. Raston. 2006. Green chemistry and the health implications of nanoparticles. *Green Chemistry* 8 (5):417–432. doi: 10.1039/B517131h.

Arora, P., S. Sharma, and S. Garg. 2002. Permeability issues in nasal drug delivery. *Drug Discovery Today* 7 (18):967–975.

Bailey, M. M. and C. J. Berkland. 2009. Nanoparticle formulations in pulmonary drug delivery. *Medicinal Research Reviews* 29 (1):196–212. doi: 10.1002/Med.20140.

Bailey, M. M., E. M. Gorman, E. J. Munson, and C. Berkland. 2008. Pure insulin nanoparticle agglomerates for pulmonary delivery. *Langmuir* 24 (23):13614–13620. doi: 10.1021/La802405p.

Behl, C. R., H. K. Pimplaskar, A. P. Sileno, J. deMeireles, and V. D. Romeo. 1998a. Effects of physicochemical properties and other factors on systemic nasal drug delivery. *Advanced Drug Delivery Reviews* 29 (1–2):89–116.

Behl, C. R., H. K. Pimplaskar, A. P. Sileno, W. J. Xia, W. J. Gries, J. C. deMeireles, and V. D. Romeo. 1998b. Optimization of systemic nasal drug delivery with pharmaceutical excipients. *Advanced Drug Delivery Reviews* 29 (1–2):117–133.

Behrens, M. and W. Meyerhof. 2011. Gustatory and extragustatory functions of mammalian taste receptors. *Physiology & Behavior* 105 (1):4–13. doi: 10.1016/j.physbeh.2011.02.010.

Bigiani, A., C. Mucignat-Caretta, G. Montani, and R. Tirindelli. 2005. Pheromone reception in mammals. *Reviews of Physiology Biochemistry and Pharmacology* 154:1–35. doi: 10.1007/s10254-004-0038-0.

Bisgaard, H., C. O'Callaghan, and G. C. Smaldone. 2007. *Drug Delivery to the Lung, Lung Biology in Health and Disease.* New York: Informa Healthcare USA, Inc.

Bitonti, A. J. and J. A. Dumont. 2006. Pulmonary administration of therapeutic proteins using an immunoglobulin transport pathway. *Advanced Drug Delivery Reviews* 58 (9–10):1106–1118. doi: 10.1016/j.addr.2006.07.015.

Broaddus, V. C., J. I. Everitt, B. Black, and A. B. Kane. 2011. Non-neoplastic and neoplastic pleural endpoints following fiber exposure. *Journal of Toxicology and Environmental Health—Part B—Critical Reviews* 14 (1–4):153–178. doi: 10.1080/10937404.2011.556049.

Bur, M., H. Huwer, C. M. Lehr, N. Hagen, M. Guldbrandt, K. J. Kim, and C. Ehrhardt. 2006. Assessment of transport rates of proteins and peptides across primary human alveolar epithelial cell monolayers. *European Journal of Pharmaceutical Sciences* 28 (3):196–203. doi: 10.1016/j.ejps.2006.02.002.

Chattopadhyay, P., B. Y. Shekunov, D. Yim, D. Cipolla, B. Boyd, and S. Farr. 2007. Production of solid lipid nanoparticle suspensions using supercritical fluid extraction of emulsions (SFEE) for pulmonary delivery using the AERx system. *Advanced Drug Delivery Reviews* 59 (6):444–453. doi: 10.1016/j.addr.2007.04.010.

Chattopadhyay, S., L. B. Modesto-Lopez, C. Venkataraman, and P. Biswas. 2010. Size distribution and morphology of liposome aerosols generated by two methodologies. *Aerosol Science and Technology* 44 (11):972–982. doi: 10.1080/02786826. 2010.498797.

Choi, H. S., Y. Ashitate, J. H. Lee, S. H. Kim, A. Matsui, N. Insin, M. G. Bawendi, M. Semmler-Behnke, J. V. Frangioni, and A. Tsuda. 2010. Rapid translocation of nanoparticles from the lung airspaces to the body. *Nature Biotechnology* 28 (12):1300–1303. doi: 10.1038/Nbt.1696.

Cotes, J. E., J. M. Dabbs, A. M. Hall, C. Heywood, and K. M. Laurence. 1979. Sitting height, fat-free mass and body fat as reference variables for lung function in healthy British children: Comparison with stature. *Annals of Human Biology* 6 (4):307–314.

Donaldson, K., F. A. Murphy, R. Duffin, and C. A. Poland. 2010. Asbestos, carbon nanotubes and the pleural mesothelium: A review of the hypothesis regarding the role of long fibre retention in the parietal pleura, inflammation and mesothelioma. *Particle and Fibre Toxicology* 7:5. doi: 10.1186/1743-8977-7-5.

Donovan, M. D. and Y. Huang. 1998. Large molecule and particulate uptake in the nasal cavity: The effect of size on nasal absorption. *Advanced Drug Delivery Reviews* 29 (1–2):147–155.

Elder, A., R. Gelein, V. Silva, T. Feikert, L. Opanashuk, J. Carter, R. Potter, A. Maynard, J. Finkelstein, and G. Oberdorster. 2006. Translocation of inhaled ultrafine manganese oxide particles to the central nervous system. *Environmental Health Perspectives* 114 (8):1172–1178. doi: 10.1289/Ehp.9030.

Fechter, L. D., D. L. Johnson, and R. A. Lynch. 2002. The relationship of particle size to olfactory nerve uptake of a non-soluble form of manganese into brain. *Neurotoxicology* 23 (2):177–183. pii: S0161-813x(02)00013-X, doi: 10.1016/S0161-813x(02)00013-X.

Foster, W. M., E. G. Langenback, and E. H. Bergofsky. 1982. Lung mucociliary function in man—Interdependence of bronchial and tracheal mucus transport velocities with lung clearance in bronchial-asthma and healthy-subjects. *Annals of Occupational Hygiene* 26 (1–4):227. doi: 10.1093/annhyg/26.2.227.

Fujita, H. and T. Fujita. 1992. *Textbook of Histology: Part 2*, 3rd edn. Tokyo, Japan: Igaku-Shoin Ltd.

Furuyama, A., S. Kanno, T. Kobayashi, and S. Hirano. 2009. Extrapulmonary translocation of intratracheally instilled fine and ultrafine particles via direct and alveolar macrophage-associated routes. *Archives of Toxicology* 83 (5):429–437. doi: 10.1007/s00204-008-0371-1.

Geiser, M., M. Casaulta, B. Kupferschmid, H. Schulz, M. Semirriler-Behinke, and W. Kreyling. 2008. The role of macrophages in the clearance of inhaled ultrafine titanium dioxide particles. *American Journal of Respiratory Cell and Molecular Biology* 38 (3):371–376. doi: 10.1165/rcmb.2007-0138OC.

Gloriam, D. E., R. Fredriksson, and H. B. Schioth. 2007. The G protein-coupled receptor subset of the rat genome. *BMC Genomics* 8:338. doi: 10.1186/1471-2164-8-338.

Grainger, C. I., L. L. Greenwell, G. P. Martin, and B. Forbes. 2009. The permeability of large molecular weight solutes following particle delivery to air-interfaced cells that model the respiratory mucosa. *European Journal of Pharmaceutics and Biopharmaceutics* 71 (2):318–324. doi: 10.1016/j.ejpb.2008.09.006.

Grenha, A., B. Seijo, and C. Remunan-Lopez. 2005. Microencapsulated chitosan nanoparticles for lung protein delivery. *European Journal of Pharmaceutical Sciences* 25 (4–5):427–437. doi: 10.1016/j.ejps.2005.04.009.

Harkema, J. R., S. A. Carey, and J. G. Wagner. 2006. The nose revisited: A brief review of the comparative structure, function, and toxicologic pathology of the nasal epithelium. *Toxicologic Pathology* 34 (3):252–269. doi: 10.1080/01926230600713475.

Heyder, J., J. Gebhart, G. Rudolf, C. F. Schiller, and W. Stahlhofen. 1986. Deposition of particles in the human respiratory-tract in the size range 0.005–15 µM. *Journal of Aerosol Science* 17 (5):811–825.

Hofmann, W., R. Winkler-Heil, and J. McAughey. 2009. Regional lung deposition of aged and diluted sidestream tobacco smoke. *Inhaled Particles X* 151:012020. doi: 10.1088/1742-6596/151/1/012020.

Hunninghake, G. W., J. E. Gadek, O. Kawanami, V. J. Ferrans, and R. G. Crystal. 1979. Inflammatory and immune processes in the human-lung in health and disease—Evaluation by bronchoalveolar lavage. *American Journal of Pathology* 97 (1):149–205.

ICRP. 1994. ICRP publication 66. Human respiratory tract model for radiological protection. ICRP Annual Report, Vol. 24 (1–3).

Jemal, A., R. Siegel, J. Q. Xu, and E. Ward. 2010. Cancer statistics, 2010. *CA: A Cancer Journal for Clinicians* 60 (5):277–300. doi: 10.1002/caac.20073.

Klaassen, C. D. 1995. *Casarett and Doull's Toxicology: The Basic Science of Poisons*, 5th edn. McGraw-Hill, New York.

Klaine, S. J., P. J. J. Alvarez, G. E. Batley, T. F. Fernandes, R. D. Handy, D. Y. Lyon, S. Mahendra, M. J. McLaughlin, and J. R. Lead. 2008. Nanomaterials in the environment: Behavior, fate, bioavailability, and effects. *Environmental Toxicology and Chemistry* 27 (9):1825–1851.

Krasteva, G. and W. Kummer. 2012. "Tasting" the airway lining fluid. *Histochemistry and Cell Biology* 138 (3):365–383. doi: 10.1007/s00418-012-0993-5.

Kuhn, R. J. 2001. Formulation of aerosolized therapeutics. *Chest* 120 (3):94s–98s.

Lai, S. K., D. E. O'Hanlon, S. Harrold, S. T. Man, Y. Y. Wang, R. Cone, and J. Hanes. 2007. Rapid transport of large polymeric nanoparticles in fresh undiluted human mucus. *Proceedings of the National Academy of Sciences of the United States of America* 104 (5):1482–1487. doi: 10.1073/pnas.0608611104.

Lay, J. C., M. R. Stang, P. E. Fisher, J. R. Yankaskas, and W. D. Bennett. 2003. Airway retention of materials of different solubility following local intrabronchial deposition in dogs. *Journal of Aerosol Medicine: Deposition Clearance and Effects in the Lung* 16 (2):153–166.

Levy, M. N., B. M. Stanton, and B. A. Koeppen. 2005. *Berne & Levy Principles of Physiology*, 4th edn. Philadelphia, PA: Elsevier Mosby.

Li, Z. and C. Kleinstreuer. 2011. Airflow analysis in the alveolar region using the lattice-Boltzmann method. *Medical & Biological Engineering & Computing* 49 (4):441–451. doi: 10.1007/s11517-011-0743-1.

Lombry, C., D. A. Edwards, V. Preat, and R. Vanbever. 2004. Alveolar macrophages are a primary barrier to pulmonary absorption of macromolecules. *American Journal of Physiology: Lung Cellular and Molecular Physiology* 286 (5):L1002–L1008. doi: 10.1152/ajplung.00260.2003.

Marquardt, H., S. G. Schafer, R. O. McClellan, and F. Welsch. 1999. *Toxicology*. San Diego, CA: Academic Press.

Matsukawa, Y., V. H. L. Lee, E. D. Crandall, and K. J. Kim. 1997. Size dependent dextran transport across rat alveolar epithelial cell monolayers. *Journal of Pharmaceutical Sciences* 86 (3):305–309.

McAughey, J., T. Adam, C. McGrath, C. Mocker, and R. Zimmermann. 2009. Simultaneous on-line size and chemical analysis of gas phase and particulate phase of mainstream tobacco smoke. *Inhaled Particles X* 151:012017. doi: 10.1088/1742-6596/151/1/012017.

Mescher, A. L. 2010. *Junqueira's Basic Histology*, 12th edn. New York: McGraw-Hill.

Miller, F. J., R. R. Mercer, and J. D. Crapo. 1993. Lower respiratory-tract structure of laboratory-animals and humans—Dosimetry implications. *Aerosol Science and Technology* 18 (3):257–271. doi: 10.1080/02786829308959603.

Moller, W., K. Felten, K. Sommerer, G. Scheuch, G. Meyer, P. Meyer, K. Haussinger, and W. G. Kreyling. 2008. Deposition, retention, and translocation of ultrafine particles from the central airways and lung periphery. *American Journal of Respiratory and Critical Care Medicine* 177 (4):426–432. doi: 10.1164/rccm.200602-301OC.

Moller, W., K. Haussinger, R. Winkler-Heil, W. Stahlhofen, T. Meyer, W. Hofmann, and J. Heyder. 2004. Mucociliary and long-term particle clearance in the airways of healthy nonsmoker subjects. *Journal of Applied Physiology* 97 (6):2200–2206. doi: 10.1152/japplphysiol.00970.2003.

Morishita, M. and K. Park. 2010. *Biodrug Delivery Systems: Fundamentals, Applications and Clinical Development*. New York: Informa Healthcare USA, Inc.

Mucignat-Caretta, C. 2010. The rodent accessory olfactory system. *Journal of Comparative Physiology A: Neuroethology Sensory Neural and Behavioral Physiology* 196 (10):767–777. doi: 10.1007/s00359-010-0555-z.

Mygind, N. and R. Dahl. 1998. Anatomy, physiology and function of the nasal cavities in health and disease. *Advanced Drug Delivery Reviews* 29 (1–2):3–12.

Nagai, H., Y. Okazaki, S. H. Chew, N. Misawa, Y. Yamashita, S. Akatsuka, T. Ishihara et al. 2011. Diameter and rigidity of multiwalled carbon nanotubes are critical factors in mesothelial injury and carcinogenesis. *Proceedings of the National Academy of Sciences of the United States of America* 108 (49):E1330–E1338. doi: 10.1073/pnas.1110013108.

Nagai, H. and S. Toyokuni. 2010. Biopersistent fiber-induced inflammation and carcinogenesis: Lessons learned from asbestos toward safety of fibrous nanomaterials. *Archives of Biochemistry and Biophysics* 502 (1):1–7. doi: 10.1016/j.abb.2010.06.015.

Nagai, H. and S. Toyokuni. 2012. Differences and similarities between carbon nanotubes and asbestos fibers during mesothelial carcinogenesis: Shedding light on fiber entry mechanism. *Cancer Science* 103 (8):1378–1390. doi: 10.1111/j.1349-7006.2012.02326.x.

Narang, A. S. and R. I. Mahato. 2010. *Targeted Delivery of Small and Macromolecular Drugs*. Boca Raton, FL: CRC Press.

Nazaroff, W. W., W. Y. Hung, A. G. B. M. Sasse, and A. J. Gadgil. 1993. Predicting regional lung deposition of environmental tobacco-smoke particles. *Aerosol Science and Technology* 19 (3):243–254. doi: 10.1080/02786829308959633.

Oberdorster, G., E. Oberdorster, and J. Oberdorster. 2005. Nanotoxicology: An emerging discipline evolving from studies of ultrafine particles. *Environmental Health Perspectives* 113 (7):823–839. doi: 10.1289/Ehp.7339.

Oberdorster, G., Z. Sharp, V. Atudorei, A. Elder, R. Gelein, W. Kreyling, and C. Cox. 2004. Translocation of inhaled ultrafine particles to the brain. *Inhalation Toxicology* 16 (6–7):437–445. doi: 10.1080/08958370490439597.

Osmond-McLeod, M. J., C. A. Poland, F. Murphy, L. Waddington, H. Morris, S. C. Hawkins, S. Clark, R. Aitken, M. J. McCall, and K. Donaldson. 2011. Durability and inflammogenic impact of carbon nanotubes compared with asbestos fibres. *Particle and Fibre Toxicology* 8:15. doi: 10.1186/1743-8977-8-15.

Patton, J. S. 1996. Mechanisms of macromolecule absorption by the lungs. *Advanced Drug Delivery Reviews* 19 (1):3–36.

Pirozynski, M. 2006. 100 Years of lung cancer (Retracted article. See vol. 103, p. 1244, 2009). *Respiratory Medicine* 100 (12):2073–2084. doi: 10.1016/j.rmed.2006.09.002.

Robinson, R. J. and C. P. Yu. 2001. Deposition of cigarette smoke particles in the human respiratory tract. *Aerosol Science and Technology* 34 (2):202–215. doi: 10.1080/027868201300034844.

Ross, M. H. and W. Pawlina. 2011. *Histology: A Text and Atlas*, 6th edn. Baltimore, MD: Lippincott Williams & Wilkins.

Sanvicens, N. and M. P. Marco. 2008. Multifunctional nanoparticles—Properties and prospects for their use in human medicine. *Trends in Biotechnology* 26 (8):425–433.

Semmler-Behnke, M., S. Takenaka, S. Fertsch, A. Wenk, J. Seitz, P. Mayer, G. Oberdorster, and W. G. Kreyling. 2007. Efficient elimination of inhaled nanoparticles from the alveolar region: Evidence for interstitial uptake and subsequent reentrainment onto airway epithelium. *Environmental Health Perspectives* 115 (5):728–733. doi: 10.1289/Ehp.9685.

Shi, L. J., C. J. Plumley, and C. Berkland. 2007. Biodegradable nanoparticle flocculates for dry powder aerosol formulation. *Langmuir* 23 (22):10897–10901. doi: 10.1021/La70020098.

Snipes, M. B. and M. F. Clem. 1981. Retention of microspheres in the rat lung after intra-tracheal instillation. *Environmental Research* 24 (1):33–41.

Standring, S. 2008. *Gray's Anatomy*, 40th edn. London: Churchill Livingstone, Elsevier.

Sznitman, J. 2013. Respiratory microflows in the pulmonary acinus. *Journal of Biomechanics* 46 (2):284–298. doi: 10.1016/j.jbiomech.2012.10.028.

Takenaka, S., E. Karg, W. G. Kreyling, B. Lentner, W. Moller, M. Behnke-Semmler, L. Jennen et al. 2006. Distribution pattern of inhaled ultrafine gold particles in the rat lung. *Inhalation Toxicology* 18 (10):733–740. doi: 10.1080/08958370600748281.

Todoroff, J. and R. Vanbever. 2011. Fate of nanomedicines in the lungs. *Current Opinion in Colloid & Interface Science* 16 (3):246–254. doi: 10.1016/j.cocis.2011.03.001.

Touw, D. J., R. W. Brimicombe, M. E. Hodson, H. G. M. Heijerman, and W. Bakker. 1995. Inhalation of antibiotics in cystic-fibrosis. *European Respiratory Journal* 8 (9):1594–1604.

Ugwoke, M. I., N. Verbeke, and R. Kinget. 2001. The biopharmaceutical aspects of nasal mucoadhesive drug delivery. *Journal of Pharmacy and Pharmacology* 53 (1):3–21.

Wang, Y. Y., S. K. Lai, J. S. Suk, A. Pace, R. Cone, and J. Hanes. 2008. Addressing the PEG mucoadhesivity paradox to engineer nanoparticles that "slip" through the human mucus barrier. *Angewandte Chemie—International Edition* 47 (50): 9726–9729. doi: 10.1002/anie.200803526.

Yacobi, N. R., L. DeMaio, J. S. Xie, S. F. Hamm-Alvarez, Z. Borok, K. J. Kim, and E. D. Crandall. 2008. Polystyrene nanoparticle trafficking across alveolar epithelium. *Nanomedicine–Nanotechnology Biology and Medicine* 4 (2):139–145. doi: 10.1016/j.nano.2008.02.002.

Yacobi, N. R., N. Malmstadt, F. Fazlollahi, L. DeMaio, R. Marchelletta, S. F. Hamm-Alvarez, Z. Borok, K. J. Kim, and E. D. Crandall. 2010. Mechanisms of alveolar epithelial translocation of a defined population of nanoparticles. *American Journal of Respiratory Cell and Molecular Biology* 42 (5):604–614. doi: 10.1165/rcmb.2009-0138OC.

Yamada, Y. 1994. *Modern Textbook of Histology*, 3rd edn. Tokyo, Japan: Kanehara & Co., Ltd.

Yamamoto, H., W. Hoshina, H. Kurashima, H. Takeuchi, Y. Kawashima, T. Yokoyama, and H. Tsujimoto. 2007. Engineering of poly(DL-lactic-co-glycolic acid) nanocomposite particles for dry powder inhalation dosage forms of insulin with the spray-fluidized bed granulating system. *Advanced Powder Technology* 18 (2): 215–228. doi: 10.1163/156855207780208592.

Yokosuka, M. 2012. Histological properties of the glomerular layer in the mouse accessory olfactory bulb. *Experimental Animals* 61 (1):13–24.

4

Buccal Route and Ingestion

Nanoproducts administered via the buccal route and ingestion are extensively utilized as nanomedicines, nanofoods, nanonutriceuticals, and nanodevices for food packaging. The buccal tissue and the gastrointestinal (GI) tract routinely undergo interactions with nanoscale substances through the process of food absorption. Food contents are mixed, crushed, degraded, and reduced to micro- or nanosized particles and low-molecular-weight (LMW) compounds in the oral cavity and are then directly absorbed by the buccal tissue. Physically degraded food contents are also chemically degraded by enzymes and microflora in the oral cavity, esophagus, stomach, and intestines, followed by further reduction to micro-/nanosized particles and LMW compounds and absorption via the GI tract.

Conventional drugs are liberated by the release of active ingredients and excipients in the GI tract, while nanomedicines are liberated by (1) the release of the microsized vehicle and excipients, (2) the release of the nanosized drug and excipients, and (3) the release of the active drug and platform.

This chapter addresses the anatomy and physiology of buccal tissue and the GI tract, in addition to practical applications, disposition, delivery systems, and endocytosis of nanoproducts administered via the buccal route and ingestion. This chapter also discusses the nanoscale world as it pertains to nanofoods and nanodevices for food packaging, as well as anatomical pores and permeability barriers relevant to the buccal route and ingestion.

4.1 Anatomy and Physiology

4.1.1 Oral Cavity and Pharynx

The oral cavity participates in the mastication and lubrication of food and the initial phase of ingestion and consists of the tongue, inner cheek, gingiva, soft palate, hard palate, and floor of the mouth (Figure 4.1a). The most important regions for the transfer of substances to systemic tissues are the sublingual mucosa in the floor of the mouth and the buccal mucosa in the cheek (Kurosaki and Kimura, 2000). The oral epithelium covers the oral mucous membranes, prevents fluid loss in the oral cavity, and protects tissues from

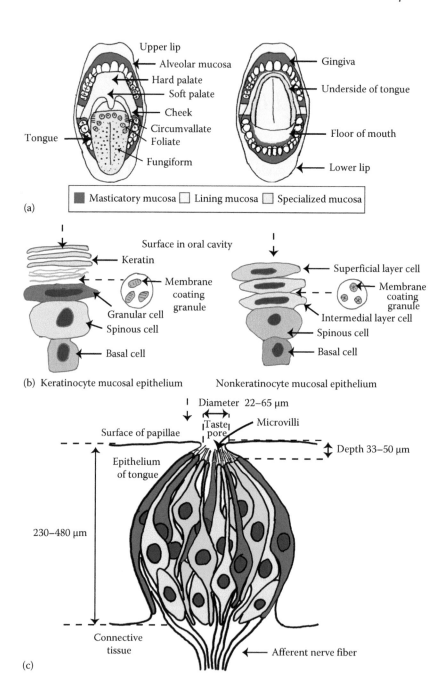

FIGURE 4.1
(a) Oral cavity, (b) mucosal epithelium, and (c) taste buds. (a: Modified from Squier, C.A. and Kremer, M.J., *J. Natl. Cancer Inst. Monogr.*, 29, 7, 2001; b: Modified from Squier, C.A. and Kremer, M.J., *J. Natl. Cancer Inst. Monogr.*, 29, 7, 2001; c: Modified from Behrens, M. et al., *Angew. Chem. Int. Ed.*, 50(10), 2220, 2011, doi: 10.1002/anie.201002094.)

toxic substances. Taste receptor–expressing cells are located in the tongue for the gustatory sensation of food.

The oral mucosa lines the surface of the oral cavity and includes the masticatory mucosa, the lining mucosa, and the specialized mucosa (Figure 4.1b) (Kurosaki and Kimura, 2000). The masticatory mucosa comprises connective tissue and keratinocytes and protects the hard palate and the gingiva from physical stress incurred during the chewing of food. The lining mucosa contains nonkeratinous cells and is found in the soft palate, the inside of the cheek, the bottom of the tongue, and the floor of the mouth (Squier and Kremer, 2001). The specialized mucosa comprises two different tissues, one keratinized and the other nonkeratinized, and is found on the upper surface of the tongue.

Intermedial layer cells in the nonkeratinized oral epithelium release membrane-coating granules (MCGs) to the extracellular matrix, where they contribute to a lipid-impermeable barrier. A demonstration of permeability constants (K_p values) in human subjects was carried out in the abdominal skin and four regions of the oral mucosa by using tritium-labeled water (Lesch et al., 1989). All four mucosal regions showed higher permeability than the skin, with K_p values of 470, 579, 772, and 973×10^{-7} in the palate, buccal mucosa, lateral border of the tongue, and floor of the mouth, respectively, versus 44×10^{-7} in the skin.

The taste-sensory tissue of a human tongue has 9–10 circumvallate papillae on average, varying from 8 to 12 or 6 to 12, depending on the source, 2 foliate papillae, and 300 fungiform papillae (Ross and Pawlina, 2011; Mescher, 2010). Each circumvallate, foliate, and fungiform papilla boasts 2200, 1300, and 1100 onion-shaped taste buds, respectively. One taste bud includes 50–100 gustatory cells (Miller, 1989). According to a study concerning fungiform papillae (Srur et al., 2010), the diameters of the taste bud and the taste pore range from 230 to 480 μm and 22 to 65 μm, respectively, while the depth of the taste pore ranges from 33 to 50 μm (Figure 4.1c). Another study estimated the diameter of the taste pore as ~20–30 μm (Segovia et al., 2002).

Taste buds are composed of types I, II, III, IV, and V cells and basal cells (Behrens et al., 2011). Types I, II, and III cells are employed for the sense of taste (Chaudhari and Roper, 2010; Iwatsuki et al., 2012; Kinnamon, 2012). Type I cells are particularly important in the sensory impression of salty flavors and contain apical slender microvilli that extend into the taste pore. These cells have a glial-like function and express ectonucleoside triphosphate diphosphohydrolase 2 and the glutamate/aspartate transporter for the excitatory neurotransmitter L-glutamate. Some type I cells also express a functional epithelial sodium channel, which is thought to be involved in the signal transduction of salty tastes.

Type II cells, also termed receptor cells, are essential for the recognition of sweet, bitter, umami (savory), and kokumi (hearty) tastes. Type II cells possess several blunt microvilli that extend into the taste pore and express two seven-transmembrane G protein–coupled receptors (GPCRs) included in the family of taste 1 receptors (T1Rs) and taste 2 receptors (T2Rs). T1Rs exist as T1R1, T1R2, and T1R3 receptors, all of which function only in the dimer form. Receptors composed of T1R1 plus T1R3 monomers respond to the "umami" taste, proposed by Ikeda (2002) to correspond to the fifth basic taste of "savory meatiness." Receptors composed of T1R2 plus T1R3 monomers respond to the sweet taste, and those composed of two T2R monomers respond to the bitter taste. Twenty-five different T2R-type receptors are documented that respond to an assortment of bitter stimuli (Meyerhof et al., 2010).

The main function of the GI tract is to digest and absorb nutritional contents and water from food. The GI tract also protects the body from exposure to harmful substances, reflected by its high content of type II cells with receptors for bitter as opposed to receptors for sweet and umami flavors. Type II cells also express other GPCRs, such as the GPCR40 and GPCR120, as well as the calcium-sensing receptor (CaSR). GPCR40 and GPCR120 mediate responses to fatty acid chains, while the CaSR mediates responses to "kokumi" tastes to enhance sweet, salty, and umami flavors (Ohsu et al., 2010).

Type III cells are required for the sensation of sour and carbonic acid tastes. Type III cells have a single, long, and slender microvillus and express the polycystic kidney disease 2–like 1 and 1–like 3 receptors, which respond to sour and carbonic acid tastes, respectively. Type III cells express synaptic proteins, engage in forming conventional synapses, and show presynaptic specialization. Therefore, type III cells are also referred to as presynaptic cells.

The direct permeation of sodium ions through a membrane-spanning cation channel contributes to the taste-sensation mechanism in type I cells, while the blockade of a protein-sensitive potassium channel is thought to predominate in type III cells by inhibiting the passage of LMW compounds. GPCRs in type II cells are critical for the translocation of nanomedicines after oral administration because nanoparticles and nanomedicines are endocytosed by a GPCR-dependent mechanism (Vodyanoy, 2010; Iwase and Maitani, 2011; Zhang et al., 2011).

4.1.2 Mucous Membranes and Permeability

Mucous membranes in the GI and respiratory systems are composed of epithelial and connective tissues. The epithelium in the stomach, small and large intestines, and the trachea is simple in nature, while the epithelium in the skin, cheek, esophagus, vagina, and cornea is stratified (Figure 4.2). The lining mucosa and the portion of the specialized mucosa at the entrance of the esophagus are covered by a stratified squamous epithelium, while the stomach-to-intestinal epithelium corresponds to a simple columnar epithelium. The MCGs released by granular cells in keratinized epithelia contain electron-dense lipid lamellae with an organized structure, whereas most MCGs released by intermedial layer cells in nonkeratinized epithelia consist of amorphous materials and nonlamellar-phase lipids and lack an organized

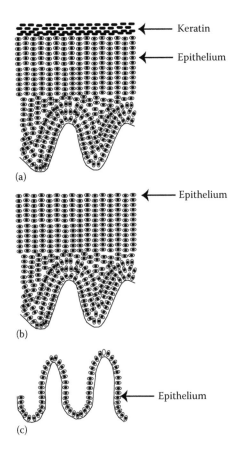

FIGURE 4.2
Epithelium of the (a) skin, (b) buccal mucosa, and (c) small intestine. (Modified from Nicolazzo, J.A. et al., *J. Control. Rel.*, 105(1–2), 1, 2005, doi: 10.1016/j.jconrel.2005.01.024.)

structure. Furthermore, MCGs are less abundant in nonkeratinized versus keratinized epithelia (Nicolazzo et al., 2005). Both types of epithelia set up chemical and physical absorption barriers that affect the membrane permeance of nanoparticles.

Chemical absorption barriers are created by metabolic enzymes and proteins secreted from the epithelial cells of the GI tract. The chemical absorption barrier primarily reflects the physicochemical properties of GI liquid and irreversible removal by first-pass organs, including the intestines, liver, and lungs (Pang, 2003). In addition, chemical absorption barriers affect drug liberation. The physical absorption barrier corresponds to an intracellular barrier formed by cellular membranes and a paracellular barrier formed by tight junctions. Improved bioavailability of nanomedicines requires the avoidance of these cellular barriers or the targeted modification of the nanoparticle surface.

4.1.3 Drug Liberation by Saliva

Saliva serves many purposes in the digestive tract and esophageal physiology (Herrera et al., 1988). The majority (90%) of the saliva is secreted by the parotid, submandibular, and sublingual salivary glands in the mucosa of the oral cavity. Saliva is a watery fluid rich in electrolytes, mucopolysaccharides, and mixed serous/mucous secretions. The remaining 10% of the saliva is secreted by small palatine, buccal, and lingual salivary glands, which secrete 70% of the mucins (heavily glycosylated proteins containing mucopolysaccharides). The average daily output of saliva in healthy human subjects who are eating normally is estimated to vary from 1000 to 1500 mL/day, while that of fasting subjects is only 500 mL/day.

The pH of saliva before secretion in the oral cavity is slightly acidic, in the range of 5.45–6.06. Saliva becomes more alkaline when salivary flow is stimulated, reaching a maximum of pH 7.8 due to its increased bicarbonate content. The acid-neutralizing capacity of saliva contributes to acid clearance in the esophagus. Saliva includes a number of enzymes (amylase, lipase, lysozymes, and peroxidases) in addition to lactoferrin, immunoglobulins, glycoproteins, and epidermal growth factor. Amylase in saliva almost always corresponds to α-amylase and is capable of digesting starch molecules into smaller polysaccharides, facilitating subsequent digestion by pancreatic amylase in the intestines. Amylase remains active until the salivary contents reach the acidic environment of the stomach.

Salivary lipase, or lingual lipase, is secreted by von Ebner's glands, located at the base of the tongue beneath the circumvallate papillae. Salivary lipase acts only on triglycerides in the acid milieu of the stomach. Because the enzyme is resistant to pepsin and its optimal pH is 3.0–6.0, salivary lipase demonstrates continued activity in the stomach. Normally, 10%–30% of dietary fat is hydrolyzed in the stomach by lingual lipase.

4.1.4 Ingestion

Nutrients are broken down in the stomach by hydrochloric acid released from parietal cells, in addition to pepsin converted from pepsinogen and secreted by chief cells. The secreted stomach fluids affect nanomedicine liberation due to the degradation of the drug formulation. The mucosal surface of the stomach is sectioned into 5–7 mm polygonal gastric areas, each of which contains numerous gastric pits (Yamada, 1994). Mucosal epithelial cells have many functions in humans and act as (1) mucus-secreting cells; (2) enteroendocrine cells in the cardiac region, surface mucosa, and mucous neck cells; (3) parietal cells; (4) chief cells; and (5) neuroendocrine cells in the fundus of the stomach and other parts of the body.

The epithelium of the small intestine (duodenum, jejunum, and ileum) contains circular folds (plicae circulares) known as the valves of Kerckring. These folds are covered with intestinal villi ~1 mm in length (Figure 4.3).

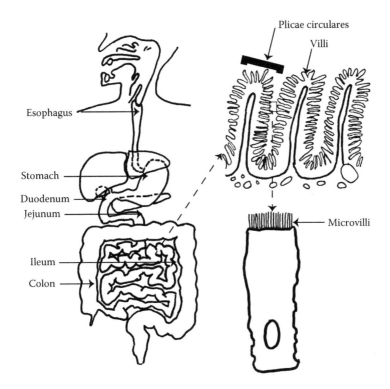

FIGURE 4.3
Plicae circulares, villi, and microvilli in the intestine.

The mucosal surface of the small intestine is composed of columnar entero-cytes (also called small intestine absorptive cells), goblet cells, paneth cells, M cells, and enteroneuroendocrine cells. The epithelium of the jejunum lacks glands at the bottom of the lamina propria and exhibits thinner plicae circu-lares, a larger number of paneth cells, and fewer enteroneuroendocrine cells than the duodenum.

The epithelium of the large intestine (colon and rectum) lacks plicae circu-lares and intestinal villi. Similar to the small intestine, the mucosal surface of the large intestine comprises columnar colonocytes/large intestine absorptive cells, goblet cells, M cells, and enteroneuroendocrine cells. The enteroneuro-endocrine cells are divided into enterochromaffin (EC) and non-EC cells. EC cells in turn comprise serotonin-releasing EC cells and somatostatin-releasing D cells, while non-EC cells comprise gastrin-secreting G cells located in the stomach and duodenum, secretin-releasing S cells, cholecystokinin-releasing I cells, M cells capable of absorbing high-molecular-weight (HMW) sub-stances, enteroglucagon-releasing L cells, glucagon-releasing A cells, and histamine-releasing EC cells (Fujita and Fujita, 1992).

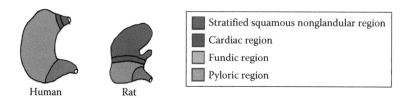

FIGURE 4.4
Anatomy of the stomach. (Modified from Kararli, T.T., *Biopharmaceut. Drug Dispos.*, 16(5), 351, 1995.)

4.1.5 Species-Specific Differences in Stomach Anatomy

The stomach displays species-specific anatomical differences in the percentage of the total area occupied by the cardiac region, fundic region, pyloric region, and stratified squamous nonglandular region (Figure 4.4) (Kararli, 1995). For example, the fundic region is wider in humans than in rats.

4.1.6 M Cells in the Intestinal Epithelium

Mucosal-associated lymphoid tissue is made up of immunoreactive cells and lymph tissue and is located around the systemic mucosa (Figure 4.5). Analogously, gut-associated lymphoid tissue (GALT) functions as an immunologic barrier in the small intestine, appendix, colon, and rectum and is located in aggregated lymphatic nodules or Peyer's patches. GALT is the largest of the lymphatic tissues and accounts for 70% of systemic lymphatic tissue. M cells are located in the GALT of the follicle-associated epithelium (FAE).

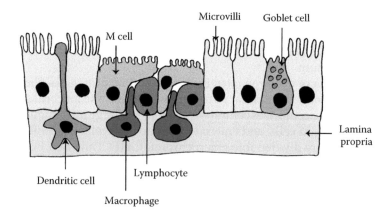

FIGURE 4.5
Overview of M cell location within the FAE in a Peyer's patch. (Modified from Corr, S.C. et al., *FEMS Immunol. Med. Microbiol.*, 52, 2, 2008.)

M cells were originally discovered by Professor Kenzaburou Kumagai in 1922 based on the uptake of heat-killed tuberculosis bacterium by the dome epithelium and the uptake of assorted antigens by Peyer's patches. The function of M cells was subsequently clarified by electron microscopic observations (Corr et al., 2008). M cells have few microvilli and lack enzyme activity, and their main function is to facilitate transepithelial transport rather than engulfment or absorption of foreign substances. M cells can import latex beads, carbon particles, liposomes, and particulate matter, including ferritin, cholera toxin–bound substances, lectins, and antiviral antigens. M cells can also import microorganisms (i.e., *Vibrio cholerae* and *Salmonella typhimurium*).

M cell–mediated import is very efficient and rapid, and the continuous process of endocytosis followed by exocytosis of transported substances occurs within 10 min (Kraehenbuhl and Neutra, 1992). The mechanism of particle uptake by M cells changes depending on particle size, surface charge, hydrophobicity, and local surface pH, as well as the presence or absence and concentration of specific receptors targeting M cells (des Rieux et al., 2005).

M cells occupy 10%–50% of the epithelium in nonhuman animals versus 5% in humans (Giannasca et al., 1999). Hence, M cells play a smaller role in humans than in nonhuman animals. However, uptake by M cells is still essential for the transport of nanoparticles and nanomedicines into systemic tissues in humans.

4.1.7 Protection against Foreign Substances by Neuroendocrine Cells

Serotonin release from enteroneuroendocrine cells is required for the control of GI motility and the clearance of toxic substances in humans. Nasal mucosal olfactory cells also release serotonin in response to odorant stimuli, including spices, fragrances, detergents, and cosmetics. Serotonin release in the intestines is linked with vomiting, diarrhea, and irritable bowel syndrome (Braun et al., 2007). Interestingly, GPCRs (T1R and T2R) and signal transduction pathways are shared by enteroneuroendocrine cells in the sensation of noxious stimuli and by type II gustatory cells in the sensation of sweet and umami stimuli. In addition, gustatory and enteroendocrine cells release some of the same hormones (e.g., cholecystokinin and glucagon-like peptide 1) (Iwatsuki et al., 2012).

Solitary chemosensory cells or brush cells in the nasal epithelium, upper larynx, trachea, bronchi, and alveoli confer protection against chemical or bacterial stimuli and inflammation. Like enteroneuroendocrine cells, chemosensory cells utilize the same receptors and signaling pathways as type II gustatory cells, except for transient receptor potential cation channel subfamily M member 5, which is only found in type II cells (Tizzano et al., 2010). Similarly, ciliated tracheobronchiolar epithelial cells utilize a T2R-like receptor to detect and eliminate harmful substances from the airway via augmented ciliary action. T2R-type receptors are involved in the sensation of bitter stimuli by type II cells (Tizzano et al., 2011), and type II and ciliated epithelial cells eliminate foreign toxic particles by preventing digestion or inhalation, respectively (Kinnamon, 2012). Finally, bitter taste receptors are found in airway smooth muscle cells (Deshpande et al., 2010).

Taken together, the results of these studies indicate that GPCRs and downstream signaling events are crucial for the protection of the body against foreign substances.

4.2 Factors Affecting Nanoparticle Disposition

The disposition of nanoparticles is influenced by the absorptive area of the GI tract, the residence time in the GI tract, and the fluid volume/enzyme release from the GI tract. Additional factors include the species and quantity of intestinal microflora, as well as pH changes brought about by secretory cells and microflora in the GI tract. I discuss each of these in detail as follows:

1. Absorptive area of the GI tract

   The absorptive area of the GI tract is the region where nanoparticles and other substances interact with cells. In an adult human, the absorptive area of the small intestine is ~120 m^2 and comprises the plicae circulares, villi, and microvilli. By comparison, the absorptive areas of the esophagus, stomach, and large intestine are ~0.02, 0.1, and 0.3 m^2, respectively (Rouge et al., 1996). Therefore, the small intestine has the largest absorptive area for interaction with nanoparticles. Nanosized substances are imported via M cells and neuroendocrine cells, while nanoparticle-derived LMW degradation products are imported by enterocytes/absorptive cells with a large surface area.

2. Residence time in the GI tract

   The residence time in the GI tract is affected by response time and disposition. The longest typical GI residence times are ~4–20 h in the large intestine, 1–10 h in the jejunum, and 0.25–3 h in the stomach (Chawla et al., 2003).

3. Fluid volume and enzyme release from the GI tract

   GI fluid volume and enzyme release both affect liberation by impacting nanoparticle degradation and disposition. Physiologically stable nanoproducts such as PEG-coated nanoparticles are effectively delivered to other tissues via the intestinal mucosa because the PEG coating inhibits nanoparticle adsorption to proteins and enzyme-regulated nanoparticle degradation and aggregation (Jung et al., 2000; Tobio et al., 2000; Behrens et al., 2002; Vila et al., 2002).

4. Species and quantity of intestinal microflora

   Intestinal microflora affect the liberation of nanoparticles and are abundant in human GI tract, from the stomach to the colon. The bacterial component of the wet fecal mass in humans is ~30%–40% (Stephen and Cummings, 1980), while microflora in the stomach and duodenum, jejunum, ileum, and colon number are 10, 10^5, 10^7, and 10^{11} per g, respectively (Rouge et al., 1996; Ley et al., 2006). Countless microflora are also found in the colon. The species diversity of intestinal microflora is simplest in infants, increases with age, and reaches maximal complexity in adults. The presence and biological activity of different species also depends on the type of food or drug consumed. Intestinal microflora have an extensive metabolic capacity and carry out substrate reduction, hydrolysis, and degradation.

5. Changes in pH induced by secretory cells and microflora in the GI tract

   Secretory cells and microflora can alter the pH in the GI tract and in this manner affect drug liberation. The normal pH of the GI tract is region dependent and mildly acidic (5.2–6.8) in the oral cavity, acidic (1.2–3.5) in the stomach, mildly acidic (4.6–6.0) in the duodenum, nearly neutral (6.8–7.3) in the jejunum, mildly alkaline (7.6) in the ileum, and alkaline (7.9–8.0 and 7.5–8.0) in the colon and the rectum (Chawla et al., 2003).

4.3 Practical Applications of Nanoproducts

Nanoproducts administered via the buccal route and ingestion are used as food contents, additives, supplements, nanomedicines, and nutriceuticals, as well as nanomaterials for food packaging. Nanoproducts used for food packaging are mainly composed of inorganic materials, while nanoproducts used for nanofoods are composed of both inorganic and organic materials. Organic materials include nanoemulsions, surfactant micelles, emulsion bilayers, double or multiple emulsions, reverse micelles, liposomes, polymeric micelles, and nanocrystals. The most popular nanoproducts for nanofoods are organic emulsions and milk protein nanotubes. NanoCrystal® is the predominant nanoproduct employed for nanomedicines.

4.3.1 Nanomedicine Bioavailability via the GI Tract

In general, nanomedicines show improved bioavailability compared with conventional drugs. Specifically, many conventional drugs or biologics show low bioavailability due to the physicochemical properties of the drug or the instability of protein or peptide biologics in the GI tract. However, nanomedicines loaded with nanosized drugs or biologics can avoid intestinal and hepatic metabolism, in addition to degradation by acids or enzymes in the GI tract, thereby maintaining the stability of active ingredients and permitting their delivery to target regions.

An example of nanotechnological engineering of drugs for oral ingestion is provided by the use of nanocrystal (Shegokar and Muller, 2010). Nanocrystal was developed by Elan Pharma International Ltd. (Dublin, Ireland) by using two technologies, crystallization for the preparation of a hydrosol (international patents issued, List and Sucker, 1988; Sucker and Gassmann, 1994) and reduction technology to convert the crystallized, large-scale form to nanosized crystals (U.S. patent issued, Liversidge et al., 1992). The manufacturers recommend reduction of the large-scale form to 400 nm or less for practical applications.

The nanocrystal formulation has a simple structure and is composed of active ingredients, contents for surface modification, and excipients (Figure 4.6). Nanocrystal is commercially available from Elan as Rapamune®, which employs rapamycin or sirolimus as an immunosuppressive or antirejection drug; Emend®, which incorporates aprepitant as an antiemetic agent; Tricor®, which incorporates fenofibrate for the treatment of hyperlipidemia; and Megace ES®, which employs megestrol acetate for the inhibition of cancer-associated angiogenesis.

Recently, Skyepharma (Muttenz, Switzerland) developed a cuboid-formed nanocrystal by using microfluidization technology. This formulation is sold as Triglide® and incorporates fenofibrate for the management of hyperlipidemia.

4.3.2 Naturally Occurring Nanofoods and Food Allergies

Naturally occurring foods contain carbohydrates, proteins, and lipids, and some of these (e.g., starch) have microscale or nanoscale structures. Starch granules and their degradation products form nanostructured substances. One starch particle of rice comprises 120–400 nm crystalline deposits (Figure 4.7), and each starch deposit is surrounded by a crystalline hard shell and a semicrystalline soft shell (Gallant et al., 1997). The hard and soft shells are composed of large and small blocks, respectively, and the blocks are composed of a crystalline layer and an amorphous layer. In wheat, the large and small blocks are 80–120 and 25 nm, respectively, and contain amylopectin clusters with a 10 nm width and a 9–10 nm height.

Most intentionally manufactured foods are microscale rather than nanoscale in size. The development of a food on the nanoscale may alter its ultrastructure and contribute to food allergies. A food allergen is defined as a substance that reacts with IgE antibodies and induces allergic reactions or sensations. Food allergens tend to be relatively heat and/or protease resistant.

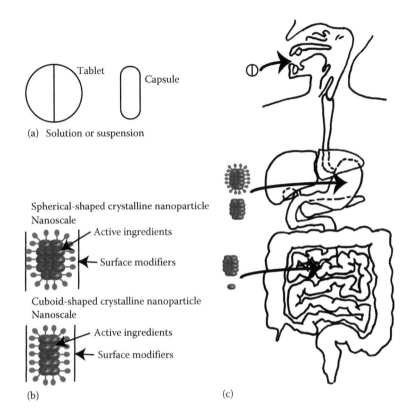

FIGURE 4.6
Nanoproducts and their degradation in the GI tract. (a) Nanocrystal in tablet, capsule, and solution form. (b) Nanocrystal with a spherical or cuboid shape. (c) Tablet degradation in the GI tract. (Modified from Shegokar, R. and Muller, R.H., *Int. J. Pharmaceut.*, 399(1–2), 129, 2010, doi: 10.1016/j.ijpharm.2010.07.044.)

The capacity of an allergen to provoke a physical immune reaction necessitates that the allergen assume an appropriate shape to be immunogenic (Aalberse, 1997). The molecular weights of the food allergens β-lactoglobulin in milk, ovotransferrin in eggs, and Ara h1 vicilin in peanuts have molecular masses of 42, 78, and 65 kDa, respectively (Aalberse, 1997). From these, the size of food allergens can be roughly estimated as 10 nm or less in diameter. In the future, it is expected that nanoscale changes will be engineered in common food allergens to render them less immunogenic.

4.3.3 Nanoproducts in Food Packaging

Conventional food packaging technology utilizes inert materials such as metal, glass, and paper. However, long-distance transportation and producer/consumer demands have led to the development of extremely sophisticated

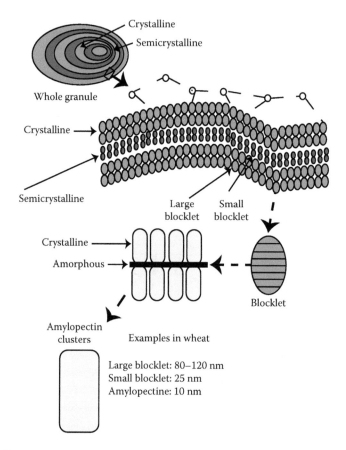

FIGURE 4.7
Structure of a starch granule. (Modified from Gallant, D.J. et al., *Carbohydr. Polym.*, 32, 177, 1997.)

new nanoproducts for food packaging, with higher function than conventional products.

Nanoproducts employed for food packaging are largely exemplified by (1) nanocomposites having exceptional strength or sealing capacity, (2) antibacterial packaging incorporating silver nanoparticles (SNPs) or SNP-containing nanocomposites, and (3) food packaging products incorporating nanoparticle sensors to monitor food freshness and to detect microbial contaminants.

4.3.4 Sealing Capacity in Food Packaging

Nanocomposites are used in food packaging to control gas exchange according to the type of food (Chaudhry et al., 2008). Plastic materials produced for carbonated beverages must be resistant to oxidation and decarbonation, while plastic materials used for fresh food must supply oxygen to increase shelf life.

Nanocomposites for gas exchange comprise a continuous phase (the filler) and a discontinuous phase. Continuous phases are generally polymeric (Figure 4.8a) and consist of polymeric starches, cellulose, polylactic acid, and proteins (Arora and Padua, 2010). Discontinuous phases are generally nanoclays. The most commonly employed nanoclay is composed of individual montmorillonite platelets (Figure 4.8a); other nanoclays are made from kaolinite, carbon nanotubes, or graphene nanosheets. Nanoclays contain hundreds of overlapping platelets. Each platelet is of the order of tens of nanometers to tens of micrometers. One platelet layer forms an octahedral monolayer with an approximate depth of 1 nm and diameter of 100–500 nm and includes aluminum, iron, or magnesium groups. The monolayer is

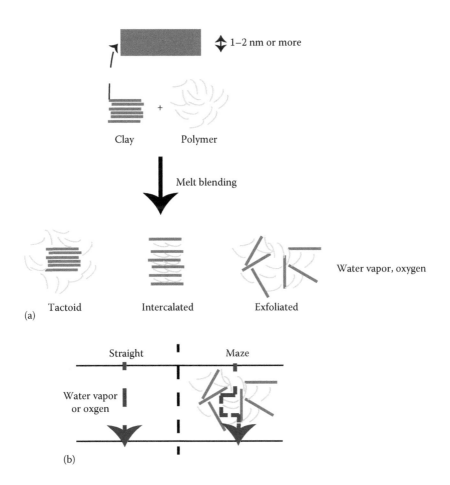

FIGURE 4.8
Manufacture of nanocomposites and maze structure. (a) Manufacture of nanocomposites. (b) Maze structure of nanocomposites. (Modified from Duncan, T.V., *J. Colloid Interf. Sci.*, 363(1), 1, 2011, doi: 10.1016/j.jcis.2011.07.017.)

wedged between tetrahedral bilayers, which includes silicon, oxygen, or hydroxyl groups (Figure 4.8a). The polymer filler in the composite surrounds the clay platelet deposit.

Polymeric nanocomposites exist as intercalated and exfoliated nanocomposites (Arora and Padua, 2010). Intercalated nanocomposites contain parallel layers of polymer filler and nanoclay deposits, and exfoliated nanocomposites contain intermingled polymer and nanoclay (Figure 4.8a).

The filling polymer interspersed among the nanoclay produces a maze structure. The pathway of tortuous movement generated by the maze structure controls the rate of gas passage (Figure 4.8b). The amount of clay used can be drastically reduced by the use of the nanocomposites versus microcomposites. When conventional microcomposites are used for food packaging, 60% more nanoclay is required for production.

4.3.5 Antibacterial Nanoproducts for Food Packaging

Antibacterial nanoproducts for the food industry incorporate two kinds of nanomaterials: SNPs and silver-containing nanocomposites. SNPs are the most common type of antibacterial nanoparticle utilized for food packaging, followed by titanium dioxide nanoparticles. Silver as a metal has more advantages than other antibacterial agents; to this end, silver is effective against more kinds of bacterial than organic antibacterial agents, including drug-resistant bacteria. Silver is highly active against methicillin-resistant *Staphylococcus aureus* and is therefore used to control hospital outbreaks (Jones et al., 2004; Gemmell et al., 2006; Echague et al., 2010). SNPs can also be easily loaded into fibers, plastic products, and polymer nanocomposites to obtain continuous antibacterial activity against numerous bacterial species. SNP-loaded plastic products are now in use as refrigerator liners (Kampmann et al., 2008), cutting boards (Kounosu and Kaneko, 2007), and food storage containers (Appendini and Hotchkiss, 2002; Quintavalla and Vicini, 2002).

Unlike silver, gold has no antibacterial activity (Kim et al., 2007). The efficacy of SNPs is due to the release of silver ions, which demonstrate antibacterial activity at low levels (Morones et al., 2005). However, SNPs are effective against silver-resistant bacteria (Eby et al., 2009). The high surface area of SNPs is thought to expedite their interactions with bacteria, subsequent ion release, and catalysis of free radical formation in bacteria; bacterial cell death then occurs via oxidative stress (Kim et al., 2007). Upon binding to bacterial membrane proteins, SNPs form dimples in the membrane, followed by the leeching of bacterial lipoproteins and loss of morphological integrity (Sondi and Salopek-Sondi, 2004). Furthermore, the SNP-provoked dimpling of the membrane leads to pore formation, enhanced membrane permeability, and cellular uptake of additional SNPs, which can alter genetic information and plasma contents.

SNPs exhibit a relatively large silver ion–covered surface area and high protein-binding effect capacity (Baker et al., 2005; Jeong et al., 2005a,b; Panacek et al., 2006; Lok et al., 2007; Martinez-Castanon et al., 2008). SNP shape influences bacterial penetration and toxicity. SNPs with a triangular shape show higher antibacterial effects against *Escherichia coli* than SNPs with a circular or rod shape, attributed to the larger surface area of the triangular shape (Pal et al., 2007). In addition, the antibacterial actions of SNPs are modified by surface charge (Lok et al., 2007), solubility, aggregation (Sondi and Salopek-Sondi, 2004; Lok et al., 2007), and surface coating (Balogh et al., 2001; Kvitek et al., 2008; Wigginton et al., 2010).

4.3.6 Nanoparticle Sensors in Food Packaging

Food freshness is usually discerned by changes in food odor or color. However, food quality can be adversely affected by circumstances outside of the consumer's control, such as long transportation times coupled with production date-dependent preservation information or the recommended expiration date (known as "shoumi kigen" or the minimum durability date, in Japan). The shoumi kigen does not always correspond to actual changes in food quality and freshness.

Food packaging materials are exposed to time-dependent gas or liquid release from foods, enabling the use of novel sensors to detect food breakdown products, harmful substances, and expiration dates. The application of nanomaterial-based chemical or electro-optical properties now permits the incorporation of nanoparticle sensors for gases, aromas, chemical pollutants, and bacterial pathogens into food packaging materials. Nanoparticle sensors can detect tiny organic molecules, gases, and bacteria and may unambiguously resolve the issue about when food has reached its final expiration date.

Nanoparticle sensors for tiny organic molecules allow the consumer to ascertain the presence of harmful substances by the naked eye. For example, cyanuric acid–coated gold nanoparticles emit red light in the absence of contaminants and blue light in the presence of even a low concentration (2.5 ppb) of the toxic substance melamine (Ai et al., 2009). In addition, although excess oxygen and liquid are important characteristics of spoiled food, conventional sensors require that the package be opened for analysis of liquid or vaporized gas contents. However, the new nanosensors are used to coat the package with a redox-active dye and can react with and alert the consumer to even very small, almost undetectable, amounts of oxygen in intact packages (Mills, 2005).

Sensors for bacteria apply the same principles as antigen–antibody complexes. The new nanosensors showed enhanced sensitivity, targeting capacity, and speed by virtue of their unique electro-optical characteristics. For example, the addition of a magnetic ferric oxide nanoparticle coated with *E. coli*–targeted antibody to *E. coli*–contaminated milk permitted the specific isolation of the bacterial contaminant (Yang et al., 2007). A conductive titanium dioxide nanowire coated with an *E. coli*–targeted antibody was used in a similar experiment. When the nanowire was connected to two gold electrodes and used to electrify targeted and nontargeted bacteria, only the targeted bacteria were affected (Duncan, 2011).

4.3.7 Nanofoods

Inorganic and organic materials are both used in the production of nanofoods and nanoscale food additives, supplements, and nutriceuticals. Silica nanoparticles, titanium dioxide nanoparticles, and metal complex-containing nanoparticles are the most common inorganic nanomaterials used in the production of nanofoods, while assorted organic materials and milk are used in the production of nanoemulsions (Figure 4.9) (Dickinson, 2011) and milk protein nanotubes (Figure 4.10) (Graveland-Bikker and de Kruif, 2006; Graveland-Bikker et al., 2006).

Food contents and food additives are characterized by their taste (sweet, salty, bitter, sour, and umami), color, flavor, and texture. New nanotechnological approaches have engendered improvements in the retention, enhancement, stability, and preservation of various food contents, but bioavailability still remains problematic. Retention and enhancement are facilitated by better protection against content release due to ingredient encapsulation, coating, and dispersion. Stability and preservation are enabled by protection against bacterial degradation via encapsulation of antioxidant compounds.

Food content retention via encapsulation in a nanocapsule is exemplified by citrus fruits, where flavor is retained over long durations of time by micelle nanotechnology (Choi et al., 2010). Furthermore, emulsion technology

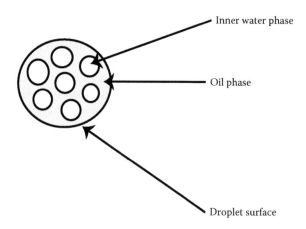

Inner water phase

Oil phase

Droplet surface

FIGURE 4.9
Nanoemulsion. (Modified from Meyer, T. et al., *J. Phys. Chem. B*, 112(5), 1420, 2008, doi: 10.1021/Jp709643h.)

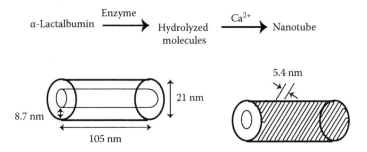

FIGURE 4.10
Milk protein nanotube. (Modified from Graveland-Bikker, J.F. et al., *Nano Letters*, 6(4), 616, 2006, doi: 10.1021/N1052205h.)

improves the delivery of functional lipids in a vast array of food products (McClements, 2010). The retention of food contents is also linked with the avoidance of unpleasant tastes or odors, for example, chitosan nanotechnology allows nanofoods to mask the odors incurred from fish oil (Klaypradit and Huang, 2008). Finally, food shapes, such as those in snack products, are maintained by silicon dioxide, magnesium oxide, and titanium dioxide nanocoatings.

It should be noted that the emulsion technology used in industrial food production generally refers to micro- rather than nanoemulsions. Nanoemulsions differ from conventional microemulsions because they are more stable against gravity (Fathi et al., 2012) and easily permit dilution by water, regardless of food content. However, most foods are not nanosized. If this were the case, nanoemulsions would be effective as flotation agents for soup. Salad dressing is an example of an emulsified product with lipid droplets suspended in water. In this case, modified starch is added for maintaining stability and viscosity.

Food content enhancement includes the reduction of food particle size for caloric reduction or health concerns with concomitant taste enhancement. For instance, excessive salt consumption is associated with the elevated occurrence of heart disease (Mohan and Campbell, 2009; Tada et al., 2011) and cancer (Inoue and Tsugane, 2005), necessitating limits on its intake (Sharp, 2004; Webster et al., 2011). However, reducing the size of salt crystals to 10–40 μm decreases the health hazards of salt by increasing salt taste intensity and reducing the amount of dietary salt employed to season foods (Charlton et al., 2007). The effect apparently stems from a size match between the reduced salt crystal size and the 20–30 μm taste pore. A similarly flavor-enhancing effect is afforded by reducing the size of lipid particles in ice cream, and mayonnaise can be generated with microscale structures of 1.5–4.4 μm (Ohashi and Shimada, 2005), 3–12 μm (Laca et al., 2010), or 11–21 μm to influence flavor characteristics (Laca et al., 2012).

A nanoscale taste-enhancing product, Slim Shake Chocolate (RBC Life Sciences, Inc., Irving, TX), was recently developed by using the

NanoCluster™ delivery system. This technology combines silica nanoparticles with an antioxidant cocoa coating (Kroese et al., 2009). The Bioral® Nanocochleate™ Delivery System (BioDelivery Sciences International, Inc., Morrisville, NC) likewise offers nanosized food particles coupled with antioxidant properties (Mozafari et al., 2006). In the latter case, a 50 nm phosphatidylserine-based carrier system was generated from GRAS contents (Zimet and Livney, 2009; Zimet et al., 2011), where the antioxidant properties effectively stabilized and preserved the food contents.

A self-assembled nanotube made from the milk protein α-lactalbumin also shows good stability. The nanotube is made by the self-assembly of α-lactalbumin following hydrolysis by *Bacillus licheniformis* and has an outer diameter of 21 nm, an inner diameter of 8.7 nm, and a unit length of 105 nm (Figure 4.10). The nanotube is used as an ingredient in children's foods (Graveland-Bikker et al., 2006, 2009). Nanotubes originating from ingestible proteins can similarly be employed as nanocapsule carriers for drugs and dietary supplements (Graveland-Bikker and de Kruif, 2006), and they have been successfully used for changing the taste or odor in soy milk, milk, yogurt, orange juice, fruit smoothies, sports drinks, coffee, Frappuccino, and other drinks.

4.3.8 Nanofoods for Health Promotion

Trans-fatty acids lengthen the preservation period of many foods and were originally developed for use in margarine or shortening at the beginning of the twentieth century. However, excessive consumption of trans-fatty acids (2 g/day) increases health risks (Stender and Dyerberg, 2004; Dhaka et al., 2011), particularly the occurrence of heart and coronary artery disease (Hunter et al., 2010; Kris-Etherton, 2010) and cancer (Vinikoor et al., 2008; Smith et al., 2009). The U.S. Food and Drug Administration (FDA) concluded in 2006 that trans-fatty acids augment the risk of heart and coronary artery disease and from that time forward required warning labels for foods containing trans-fatty acid in excess of 0.5 g per serving (Moss, 2006).

Eicosapentaenoic and docosahexaenoic acids are ω-3 fatty acids found in fish. ω-3 fatty acids are essential for normal brain development, and ω-3 fatty acid deficiency adversely affects learning and memory (Helland et al., 2003; McNamara and Carlson, 2006; Lundqvist-Persson et al., 2010). Consumption of ω-3 fatty acids may lower the risk of heart disease (Bhupathiraju and Tucker, 2011) and delay the occurrence of Alzheimer's disease (Morris et al., 2005). The U.S. National Institutes of Health concluded that a healthy adult human subject should consume ~0.65 g of ω-3 fatty acids per day (Simopoulos et al., 2000).

The low frequency of heart and coronary artery disease in Japan is ascribed to a low level of total cholesterol in the blood plasma. The Japanese diet is characterized by a decreased consumption of trans-fatty acids and salt, together with an increased consumption of ω-3 fatty acids, fish, soy products, fruits, and vegetables. This diet actually lowered the occurrence of heart disease in Japan every year between 1960 and 1980 (Tada et al., 2011).

Recently, casein- and milk protein–based nanocomposites have been engineered for the delivery of ω-3 fatty acids (Livney, 2010; Abd El-Salam and El-Shibiny, 2012). Nanocomposites composed of β-lactoglobulin and pectin and loaded with ω-3 fatty acids can protect the fatty acid from oxygen-induced degradation and extend the shelf life (Zimet and Livney, 2009; Jones et al., 2010). The composites also permit ω-3 fatty acid supplementation of home-cooked goods, including cakes, muffins, pasta noodles, soups, and cookies.

4.4 Risks and Benefits of Nanoscale Food Packaging Products and Nanofoods

Food packaging products containing nanocomposites and other nanomaterial constituents are typically inert and have no adverse effects on humans in and of themselves. Furthermore, polymer-based nanocomposites are beneficial in that they allow gas exchange, effective loading of antibacterial agents (Althues et al., 2007), and long-term antibacterial effects via the release of SNPs (Furno et al., 2004; Roe et al., 2008). Nonetheless, nanocomposites are associated with certain health risks, including the release of chemical substances from the polymer filler (Pocas and Hogg, 2007; Khaksar and Ghazi-Khansari, 2009; Silva et al., 2009; Xu et al., 2010), as well as the release of silver from the SNPs (Benn and Westerhoff, 2008; Benn et al., 2010). Furthermore, nanocomposites containing polymers can release the polymer, which becomes an issue when endocrine-disrupting compounds are included in the contents of nanofood packaging (Muncke, 2009). However, it must be noted that food packaging composed of conventional materials can also release harmful chemical substances (Arvanitoyannis and Bosnea, 2004).

The release of silver from antibacterial packaging or the direct consumption of SNPs can induce significant renal toxicity (Diamond and Zalups, 1998) and/or SNP accumulation in neural cells (Panyala et al., 2008; Luther et al., 2011). Moreover, the in vitro exposure of undifferentiated PC12 cells to 1 µM silver nitrate inhibited their neurodifferentiation into acetylcholine- and dopamine-expressing phenotypes and elicited a significant increase in lipid peroxidation (Powers et al., 2010). Another in vitro study using human hepatocellular carcinoma cells showed that SNPs adversely affected the cell cycle at low concentrations and provoked cellular morphological abnormalities, a reduction in cell number, cell shrinkage, and chromosomal disturbances at high concentrations. The toxicities were induced not only by released silver but also by intact SNPs (Kawata et al., 2009).

Cellular uptake and toxicity of SNPs is influenced by SNP size (Carlson et al., 2008). An in vitro cytotoxicity study of 15, 30, and 55 nm SNPs demonstrated that the half-maximal effective concentration (EC_{50}) was ~30 μg/mL for both 15 and 30 nm SNPs, whereas the EC_{50} of the 55 nm SNPs (>75 μg/mL) showed substantially lower toxicity. Furthermore, Tang et al. (2009) performed a comparative toxicological study after subcutaneous injection of 2–20 μm silver microparticles and 50–100 nm SNPs at a dose of 62.8 mg/kg in rats. The injection of the SNPs resulted in the specific distribution of silver to the kidney and brain, as well as blood–brain barrier destruction and astrocyte swelling in the brain, although anionic stealthy nanoparticles should hypothetically distribute to the liver or the spleen after injection (Igarashi, 2008).

SNPs are also detrimental to male fertility. Low-level exposure (10 mg/mL) to 15 nm SNPs altered cellular morphogenesis and evoked mitochondrial disorders in spermatogonial stem cells (Braydich-Stolle et al., 2005). Nonetheless, despite the recent increase in production and consumption of SNP-containing products and the associated negative press, 7–9 nm SNPs were in fact first manufactured as colloidal silver 120 years ago (Nowack et al., 2011).

The minimally effective concentration of 45–50 nm SNPs against *E. coli*, *V. cholerae*, *Shigella flexneri*, and at least one strain of *S. aureus* is 2–4 mg/mL (Sarkar et al., 2007). An FDA food additive law passed in 2009 permits silver nitrate concentrations of <17 μg to be used in commercial packaging (FDA Fed. Regist. 74, 2009).

4.5 Bioavailability of Nanomedicines and Nanofoods

4.5.1 Bioavailability of Conventional Drugs and Supplements

Delivery via the stomach and intestines is noninvasive and enables self-administration of drugs and is the route used for food consumption and the release of chemical substances from consumer products (e.g., chewable vitamins) via degradation. This is also the occupational exposure route for certain manufactured nanoproducts. A survey of drug bioavailability (Fasinu et al., 2011) indicated that 60% of the drugs marketed as medicines are administered via the oral route, and most of these show low bioavailability. For example, 90% of orally administered medicines are lost by metabolism before systemic distribution. This is significant in that low bioavailability can lead to pronounced interindividual variation in drug actions.

Orally administered protein or peptide biologics show low bioavailability via enzyme inactivation. After oral administration of insulin to rodents, bioavailability was 1% or less (Perakslis et al., 2007). Furthermore, the bioavailability of buprenorphine, a potent opioid analgesic, showed large interindividual variations in humans after sublingual and buccal administration.

The bioavailability was approximately 51% and 28% after sublingual and buccal administration of 4 mg buprenorphine, respectively, and the area under the curve (AUC) for sublingual and buccal administration ranged from 9 to 36 and 1.6 to 30 h ng/mL, respectively (Kuhlman et al., 1996). By comparison, the AUC for intravenous injection of 1.2 mg buprenorphine ranged from 11 to 22 h ng/mL.

Bioavailability for food applications is engineered to maximize absorption of food contents. Hydrophilic vitamin C is engineered as a lipophilic suspension for skin care, while lipophilic vitamin A is engineered to increase its hydrophilicity for use as a dispersed compound. In the case of vitamins, such modifications are easy to formulate. However, many drugs, foods, and supplements are difficult to alter to increase bioavailability. Although ω-3 fatty acids and various herbal supplements exert numerous therapeutic and preventative actions, their intestinal absorption is low (Averina and Kutyrev, 2011). The herbal supplement resveratrol inhibits seven subtypes of silent information regulator genes termed sirtuins (Kelly, 2010) and is widely sold and consumed as an antiaging drug. However, its bioavailability is <1% (Walle, 2011) because the drug is changed into an inactive form by intestinal and hepatic metabolism.

4.5.2 Mucous Membranes of the Oral Cavity

After drug delivery to the oral cavity, absorption occurs via the mucosa at the bottom of tongue and the inside of the cheek. The cheek mucosa consists of a stratified, nonkeratinized squamous epithelium that shows regional differences in thickness, a basal lamina, and a supporting tissue underlayer. The overall thickness of the cheek mucosa is 500 μm (~40–50 cells), while the thickness of the tongue mucosa is 100–200 μm (Harris et al., 1992). A pharmacokinetic report in the clinical study of misoprostol, an acid suppressant, showed that absorption via the tongue mucosa versus the cheek mucosa led to four- and fivefold higher drug concentrations in the blood, as assessed by the AUC and the peak concentration, respectively (Schaff et al., 2005). This result suggests that the mucosa on the bottom of the tongue confers high drug bioavailability due to its high permeability. On the other hand, the cheek mucosa is superior to the tongue mucosa in terms of surface area, high blood circulation, and turnover rate of the mucosa (namely, the time required for the replacement of the epithelium) (Squier and Nanny, 1985; Collins and Dawes, 1987; Squier and Wertz, 1993; Sohi et al., 2010).

Drug delivery via the buccal mucosa avoids first-pass hepatic and intestinal metabolism, and therefore, drugs are directly delivered to the blood for circulation. Hence, this delivery system provides an effective alternative to systemic delivery. Regardless of the presence or absence of keratinocytes, stratified squamous epithelium is characterized by numerous 100–400 nm MCGs in the epithelium and extracellular matrix. MCGs have a barrier-forming role at the epithelial surface of the skin, respiratory system, and oral mucosa. Epithelial transport studies showed that exogenously administered proteins did not pass further than epithelial layers 1–3 in a rabbit and a rat, but they traveled as far as the basal lamina in layers 3–8 in a monkey after removal of MCGs by horseradish peroxidase treatment (Tanaka, 1984; Romanowski et al., 1988). This finding suggests that entry into layer 3 is the minimal requirement for nanoparticle passage into the stratified squamous epithelium.

MCGs consist of a lamellar internal structure surrounded by a 6 nm thick trilamellar membrane. MCGs are produced in the Golgi zone, move from the middle portion to the upper surface of the epithelial cell, fuse to the uppermost cell membrane, and then exit to the intercellular space (Lavker, 1976). For this reason, epithelial permeability is enhanced by the removal of accumulated MCGs.

4.5.3 Conventional Drug Delivery via the Buccal Mucosa

Drug delivery via the buccal mucosa avoids the effects of metabolism in the intestines and the liver, as previously discussed for the delivery of conventional medicines (Hoogstraate and Wertz, 1998). The molecular cutoff size for mucosal permeability was clarified in a mucosal model of the oral cavity in a pig (Junginger et al., 1999). In this model, fluorescent dye molecules of 10 kDa passed through the mucosa, but dye molecules of 20 kDa did not.

Lipids, like MCGs, have a barrier-forming role to limit epithelial permeability and are discharged into the intercellular space along with MCGs (Squier et al., 1991). Another study of the porcine buccal mucosa showed that the keratinocyte-rich palatal epithelium and the gingival epithelium both contain high levels of ceramides and cholesterol and low levels of glycosylceramides and cholesterol esters. The reverse was true for the cheek and bottom-of-the-tongue epithelia, which contain high levels of phospholipids, cholesterol esters, and glycosylceramides and low levels of ceramides (Squier et al., 1991). These molecular differences affect region-specific permeability in buccal delivery. For instance, the lipophilic membrane on the cheek epithelial surface is normally surrounded by hydrophilic molecules and discharges polar lipids along with MCGs. Accordingly, paracellular and transcellular pathways both participate in drug delivery via the cheek mucosa (Figure 4.11). The epithelium of the small intestine is also surrounded by hydrophilic space, but the small intestine epithelium lacks polar lipids.

Although the epithelial membrane of the cheek mucosa is lipophilic and forms a barrier to the transport of hydrophilic materials, such compounds can easily travel via the paracellular pathway. Tight junctions, while abundant in the intestinal epithelium, are rare in the cheek epithelium (Barnett and Szabo, 1973; Harris and Robinson, 1992). Therefore, the main pathway

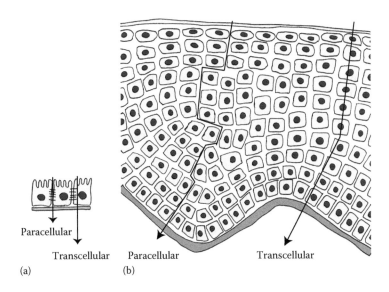

Paracellular

Transcellular   Paracellular   Transcellular

(a)   (b)

FIGURE 4.11
Epithelial transport pathways. (a) Simple columnar epithelium and (b) stratified squamous epithelium. (Modified from Nicolazzo, J.A. et al., *J. Control. Rel.*, 105(1–2), 1, 2005, doi: 10.1016/j.jconrel.2005.01.024.)

via the intestines is transcellular transport, whereas the main pathway via the cheek is passive paracellular transport through the mucosa (Lavker, 1976).

Transport-enhancing factors are well documented in paracellular drug delivery via the buccal mucosa (Nicolazzo et al., 2005); nonetheless, they are not biologically successful because they subject the oral cavity to airborne contaminants and reduce resistance against viral invasion (Song et al., 2004).

Delivery via the esophagus is difficult for absorption or bioavailability, because the contact time of drug with esophagus mucosa is 10 s at best and is only 1.4 s for drug delivery via the esophageal mucosal epithelium. However, delivery via the buccal cavity permits higher drug absorption or bioavailability by increasing the contact into buccal mucosa. For this reason, efforts are underway to lengthen contact time for buccal delivery by engineering drugs in the form of slowly dissolving tablets and lozenges, chewing gum (Rassing, 1994), oral sprays, patches, and films (Pather et al., 2008).

4.5.4 Nanomedicine Bioavailability via the Buccal Mucosa

Insulin-loaded nanomedicines, such as Oral-lyn™ (Generex Biotechnology Corp., Toronto, Canada), have recently been developed for buccal delivery. However, Oral-lyn and similar drugs are not yet available on the market

(Heinemann, 2011). Their failure is attributed to low bioavailability and consequent interindividual variation (Kalra et al., 2010).

4.5.5 Nanomedicine Bioavailability in the Intestines via Paracellular Transport

Nanomedicines can be transported in the intestines by four mechanisms: paracellular transport, transcellular transport, transcytosis, and carrier-mediated transport (Figure 4.12). Paracellular transport via the intestinal mucosa occurs via occluding junctions (i.e., zonula occludens/tight junctions) and anchoring junctions (i.e., zonula adherens or macula adherens/desmosomes). Transcellular transport allows the passive transport of LWM drugs or nanomedicines via endocytic transport. Transcytosis occurs when the proximal surface of the cell membrane is surrounded by small, fat vesicles. The small vesicle is transported into the cell, released with its contents to the underlying basal lamina, and transported into another cell as transcellular import or exocytosed. Finally, carrier-mediated transport refers to specifically targeted transport. The nanomedicine binds reversibly with a specific molecular complex in the plasma membrane lipid bilayer. The complex is imported into the cell, makes contact with the basal lamina, forms a new complex with the underlying plasma membrane, and is then exported to the exterior of the cell (Blanchette et al., 2004).

The intercellular space is narrowed by cell–cell adhesions, restricting physiological paracellular transport to 0.3–1 nm particles (Nellans, 1991). Nanomedicine delivery via the paracellular pathway in the intestines is therefore limited by cell shape and the strength of cell–cell adhesions. Chitosan

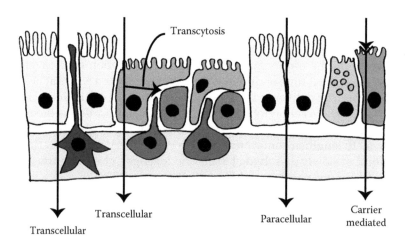

FIGURE 4.12
Transport pathways through cell layers. (Modified from Peppas, N.A. and Carr, D.A., *Chem. Eng. Sci.*, 64(22), 4553, 2009, doi: 10.1016/j.ces.2009.04.050.)

promotes the depolarization of F-actin in cells, modifying protein localization to tight junctions and inhibiting protein kinase C (Ma et al., 2005; Smith et al., 2005). Furthermore, application of thiolated polymers elongates tight junctions via inhibition of protein phosphate tyrosine (Bernkop-Schnurch et al., 2003) to permit more ready passage of medicines. Nonetheless, the loosening of tight junctions also permits the passage of otherwise excluded toxic products and biological pathogens into the GI tract and the blood (Carino and Mathiowitz, 1999).

Intestinal paracellular transport includes both M cell– and dendritic cell (DC)–mediated pathways (Didierlaurent et al., 2002). The entry of pathogenic microorganisms mainly occurs via M cells that are concentrated in the FAE overlying Peyer's patches. M cells are very selective and do not allow entry of all microorganisms. The entry of digested food antigens most likely takes place across both M cells and epithelial cells that allow entrance of particles smaller than 28 nm (Rimoldi and Rescigno, 2005).

DCs are located in intestinal mucosal tissues and especially in tissues underlying the mucosal epithelium of intestinal villi and Peyer's patches. DCs express all identified tight junction proteins, including claudin-1, occludin, and zonula occluden-1. Under normal conditions, the tight junctions between DCs and epithelial cells generate the epithelial barrier. However, during infection, DCs in lamina propria open the tight junctions between adjacent epithelial cells via process elongation and downregulation of claudin-1, occludin, and zonula occluden-1, allowing direct capture of pathogenic bacteria across the mucosal epithelium (Hofman, 2003; Rimoldi and Rescigno, 2005). Lamina propria DCs can effectively discriminate between pathogenic and nonpathogenic bacteria.

Intestinal inflammatory diseases such as Crohn's disease modify epithelial permeability, and thus, nanoparticles are more easily transported during outbreaks of intestinal inflammation (Didierlaurent et al., 2002; des Rieux et al., 2006). However, nanoparticles cannot penetrate intestinal mucosa via paracellular transport under normal physiological conditions.

4.5.6 Transcellular Transport of Nanomedicines

In transcellular transport via the intestinal mucosa, nanomedicines are endocytosed, released, or unreleased within the cell and exocytosed to the basal lamina for transcellular delivery. The uptake of particles via endocytosis is mediated by absorptive cells (enterocytes in the small intestine and colonocytes in the large intestine), M cells, and probably neuroendocrine cells. Nonspecific import comprises three mechanisms of delivery. The first delivery mechanism involves the protection of encapsulated active ingredients by high prodrug stability and initial contact with the mucosa without inactivation of the active ingredients. The second delivery mechanism involves enhancement of adhesion with the anionic intestinal mucosa by using particles coated with cationic substances and requires no special receptors.

The third delivery mechanism necessitates interaction with the mucosa by using particles coated with ligands showing affinity for or selectivity against specific membrane receptors located in intestinal absorptive cells, M cells, or neuroendocrine cells as same as carrier-mediated transport.

Critical factors affecting import by endocytosis in the intestinal epithelium and translocation into the blood are (1) particle size, (2) physicochemical properties of the particle surface, and (3) physiological properties of the GI tract. I discuss each of these in detail as follows:

1. Particle size

   Previous work showed that polystyrene latex nanoparticles were maximally endocytosed at a particle diameter of 50–100 nm. Particles of 1000 nm or more were trapped in Peyer's patches (Jani et al., 1989, 1990). Furthermore, a study of the inflamed colon mucosa in rats employed particles of 100 nm, 1 μm, and 10 μm and showed that the 100 nm particles were absorbed most readily (Lamprecht et al., 2001).

2. Physicochemical properties of the particle surface

   Charged particles demonstrate low bioavailability via the oral route because of physically repellent, repulsive forces between the particle and the mucosal liquid. Increased hydrophobicity enhances membrane penetration (Norris and Sinko, 1997) but decreases endocytosis (Hussain et al., 2001).

3. Physiological properties of the GI tract

   Bacterial invasion into the intestines upregulates the translocation of particles across the FAE covering Peyer's patches (Borghesi et al., 1999; Meynell et al., 1999; Gebert et al., 2004). Therefore, the translocation of nanomedicines is altered by changing the physiological properties of the GI tract.

4.6 Cellular Endocytosis of Nanoparticles

Orally administered nanoparticles are absorbed by three types of cells: the intestinal absorptive cells, representing a high proportion of the cells in the GI tract, the membranous epithelial cells (M cells) overlying Peyer's patches in the GALT, and the neuroendocrine cells.

4.6.1 Endocytosis via Absorptive Cells

The uptake of coated nanoparticles is limited to endocytosis for absorptive cells, which comprises macropinocytosis, clathrin- or caveolin-mediated endocytosis, and clathrin-/caveolin-independent endocytosis (Conner and

Schmid, 2003; Doherty and McMahon, 2009; Canton and Battaglia, 2012; Kettiger et al., 2013). Most pinocytic pathways (e.g., clathrin-mediated endocytosis) are receptor-mediated processes involving specific receptor–ligand interactions. Cells can import nanoparticles with a diameter of ≤150 or ≤200 nm via clathrin-coated vesicles (CCVs) via clathrin-mediated endocytosis (Rejman et al., 2004; Smith et al., 2012) and particles of ≥1 µm via macropinocytosis (Doherty and McMahon, 2009; Canton and Battaglia, 2012; Kettiger et al., 2013). The mechanism of macropinocytosis encompasses membrane ruffling by an actin-dependent process (Figure 4.13).

The impact of nanoparticle physiochemical properties such as hydrophobicity, hydrophilicity, and surface charge on uptake by absorptive cells remains controversial. Nanoparticles coated with hydrophobic versus hydrophilic substances demonstrated extensive import and reduced systemic bioavailability in an early study (Eldridge et al., 1990), but more recent in vivo work showed that PEG coating of hydrophilic particles increased their bioavailability (Tobio et al., 1998, 2000; Jung et al., 2000; Vila et al., 2002, 2004).

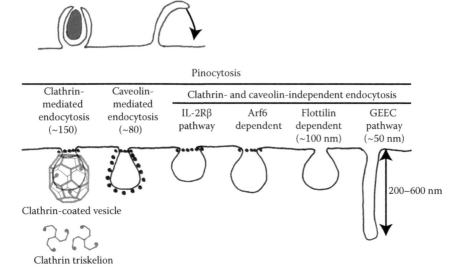

FIGURE 4.13

Multiple portals of entry into the mammalian cell. IL-2Rβ, interleukin 2 receptor β; Arf6, ADP ribosylation factor 6; and GEEC, GPI-AP-enriched early endosomal compartment. (Modified from Conner, S.D. and Schmid, S.L., *Nature*, 422(6927), 37, 2003, doi: 10.1038/nature01451; Canton, I. and Battaglia, G., *Chem. Soc. Rev.*, 41(7), 2718, 2012, doi: 10.1039/C2cs15309b; Doherty, G.J. and McMahon, H.T., *Annu. Rev. Biochem.*, 78, 857, 2009, doi: 10.1146/annurev.biochem.78.081307. 110540; Kettiger, H. et al., *Int. J. Nanomed.*, 8, 3255, 2013, doi: 10.2147/Ijn.S49770.)

Nanoparticles showing enhanced mucosal adhesion often demonstrate augmented nonspecific import by absorptive cells (Mikos et al., 1991; Smart, 2005). This phenomenon is associated with an increased residence time and partitioning out of mucosal tissue into systemic circulation (Lehr et al., 1991). The retention time at the enterocyte surface is also affected by the physiological turnover rate of the mucosal layer. In addition, nanoparticles can tightly adhere to unintended locations in the stomach mucosa or intestinal contents due to nonspecific entrapment by mucin proteins. Regardless of the mucosal layer turnover rate, an attenuated retention time by mucin attachment lowers engulfment by absorptive cells.

The most important physiological mechanism of uptake by absorptive cells is clathrin-mediated endocytosis, with a cutoff size for import of ≤200 nm (Rejman et al., 2004). The size of a CCV is dependent on the size of the cargo during bud formation. The size of a cargo inside a clathrin basket was estimated at 150 nm in a study using HeLa cells (Smith et al., 2012). CCV formation involves bud formation on the plasma membrane in the early stage of cell entry and then the formation of a clathrin cage on the outer side of the CCV after the assembly of the three-legged clathrin triskelion (each leg measures ~52 nm) (Ferguson et al., 2008). The upper size limit of a CCV was estimated to be ≤200 nm in diameter in a size-dependent internalization study that used fluorescent nano- and microspheres with sizes of 50, 100, 200, 500, and 1000 nm in diameter and a murine melanoma cell line (Rejman et al., 2004). A study of CCV formation and fluorescent nanoparticle revealed that 200 nm nanoparticles can be endocytosed. Therefore, nanoparticles of ≤150 nm can be readily imported via clathrin-mediated endocytosis, while those of >200 nm cannot.

Endocytosis via absorptive cells is effectively augmented by decorating the nanoparticle surface with targeting molecules that interact with cell surface receptors or sugars (Hussain, 2000). This approach is typified by the attachment of lectins to the nanoparticle surface, including wheat germ agglutinin (WGA), concanavalin A (ConA), *E. coli*–derived heat-labile toxin (LTB), tomato lectin (TL), and *Aleuria aurantia* fungus-derived lectin (AAL). WGA binds with *N*-acetyl-D-glucosamine and sialic acid on the cell surface (Gabor et al., 1998, 2004), ConA binds with α-D-mannose, LTB binds with GM1 ganglioside and galactose (Russell-Jones et al., 1999), TL binds with *N*-acetyl-D-glucosamine, and AAL binds with L-fucose.

WGA-coated nanoparticles target human epithelial colorectal adenocarcinoma Caco-2 cells and absorptive cells and are endocytosed via the epidermal growth factor receptor (Pusztai et al., 1993; Lochner et al., 2003). WGA-coated nanoparticles showed a 12-fold higher interaction with Caco-2 cells relative to PEG-coated nanoparticles (Gref et al., 2003). Moreover, WGA-, ConA-, and LTB-coated 50–150 nm nanoparticles all exhibited enhanced targeting activity and import into cells compared with controls (Russell-Jones et al., 1999). However, lectins have the capacity to induce inflammation and allergic reactions in the GI tract, necessitating alternative approaches to reduce their immunogenicity while retaining their binding activity (Russell-Jones et al., 1999).

4.6.2 Endocytosis via M Cells

The energy-dependent transcellular pathway predominates in the translocation of nanoparticles by M cells (Brayden et al., 2005). M cell–mediated endocytosis of nanoparticles, bacteria, and HMW substances includes three mechanisms: fluid-phase endocytosis, absorption endocytosis, and phagocytosis (Owen, 1977; Neutra et al., 1987; Florence, 2004). Nanoparticle import via M cells is affected by (1) the particle size, (2) the hydrophilicity/hydrophobicity balance of the particle coupled with surface charge, and (3) the presence of targeting molecules on the nanoparticle surface. I discuss each of these in detail as follows:

1. Particle size

    Particles of ≤1000 nm are imported by M cells and translocated to the basal lamina (Clark et al., 2001), while those of ≥5 µm are imported by M cells and trapped by Peyer's patches (Eldridge et al., 1990). The optimal size for endocytosis by M cells is 1000 nm or less (Jani et al., 1989; Gullberg et al., 2000; Clark et al., 2001) and, by some accounts, 200 nm or less (Jani et al., 1990; des Rieux et al., 2005).

2. Hydrophilicity/hydrophobicity balance and surface charge

    The solubility of nanoparticles in water and their surface charge (anionic, cationic, or neutral) hypothetically impact endocytosis by M cells, enterocytes, and neuroendocrine cells. However, because of the low bioavailability of endocytosed nanoparticles, the experimental evidence remains inconclusive. Several studies showed enhanced import of hydrophobic nanoparticles by M cells and GALT and enhanced import of hydrophilic nanoparticles by absorptive cells (LeFevre et al., 1985; Eldridge et al., 1990; Hillery and Florence, 1996; des Rieux et al., 2005). Although charged nanoparticles can also be endocytosed by M cells, their import is diminished relative to that of nonionic, hydrophobic nanoparticles. Furthermore, nanoparticle/M cell interactions are influenced by the zeta potential of the nanoparticle surface (Florence, 2005). Anionic or neutral nanoparticles showed better import via Peyer's patches than cationic nanoparticles in one study (Shakweh et al., 2005), and another study predicted that hydrophobicity plus anionic charge would provide the best combination of properties for nanoparticle endocytosis by M cells (Jung et al., 2000).

3. Presence of targeting molecules on the nanoparticle surface

    Endocytosis of nanoparticles via M cells is enabled by coating the particle surface with targeted molecules against M cells, such as lectins, Igs, GM1 ganglioside, and bacterial pathogens. Foster and Hirst (2005) demonstrated 100-fold higher binding affinity between M cells and polystyrene nanoparticles (500 nm) coated with *Ulex europaeus* agglutinin 1 lectin than between M cells and uncoated nanoparticles (Foster and Hirst, 2005). Lectins originating from *Sambucus nigra* and *Viscum album* also showed affinity for the FAE surface in humans (Jepson et al., 1996; Sharma et al., 1996). Furthermore, M cells express surface IgA receptors and interact with IgA (Roy and Varvayanis, 1987; Weltzin et al., 1989; Mantis et al., 2002), and IgA-coated latex beads are imported into M cells 20–30 times more rapidly than uncoated beads (Porta et al., 1992). Coating of nanoparticles with ganglioside GM1 (the receptor for the B subunit of cholera toxin) also successfully targeted the particles to M cells, as well as to absorptive cells (Brayden, 2001; Brayden et al., 2005).

Targeting of appropriately labeled nanoparticles to M cells can simulate interactions between the cells and pathogenic bacteria (e.g., *Yersinia*, *Salmonella*, and *Shigella*). For example, polymeric 280 nm nanoparticles were prepared after incubation of the polymer with *Salmonella enteritidis* extract. Oral administration of the nanoparticles to rats showed extensive distribution to Peyer's patches (Salman et al., 2005), indicating that molecules present in the bacterial extract targeted M cells in the absence of intact *Salmonella*.

4.6.3 Endocytosis via Neuroendocrine Cells

Caveolin- and clathrin-independent endocytosis via neuroendocrine cells is a demonstrated uptake mechanism in humans (Conner and Schmid, 2003). Clathrin-dependent endocytosis is relatively slow at $t^{1/2} > 1$ min, whereas caveolin- or clathrin-independent endocytosis is rapid, occurring at $t^{1/2} < 10$ s after stimulated secretion, and involves the neuron-specific isoform of dynamin. A detailed description of nanoparticle endocytosis via the systemic route is provided in Chapter 6.

4.7 Disposition of Nanoparticles Following Oral Administration

4.7.1 Disposition of Biodegradable Nanomedicines

True success in biodegradable drug delivery via the oral route is demonstrated by (1) improvements in the drug liberation parameter, with avoidance of biodegradation until the prodrug is delivered to target cells together with loaded active ingredients; (2) systemic circulation via blood vessel interactions with absorptive cells, M cells, and neuroendocrine cells; and (3) delivery of intact active ingredients for peptides or proteins that are easily inactivated in the blood vessels.

Subcutaneous injection of insulin is a daily complication of self-therapy for insulin-dependent diabetic patients. Insulin tablets for oral administration would decrease this burden and improve the quality of life. However, such formulations are associated with low bioavailability, as discussed in Section 4.5.4. For instance, oral administration of 300–400 nm insulin-loaded nanoparticles in a poly(isobutylcyanoacrylate) platform entered the vascular system and drastically decreased glucose levels in the blood in rats (Damge et al., 1988, 1990). Although the insulin-loaded nanoparticles were absorbed by intestinal cells, they were extensively degraded by absorptive cells and M cells (Pinto-Alphandary et al., 2003). Moreover, the intestinal barrier coupled with low bioavailability provoked high interindividual variations in insulin concentration following administration of 270–340 nm insulin-loaded nanoparticles in a chitosan platform (Cournarie et al., 2002).

4.7.2 Disposition of Nonbiodegradable Nanomaterials

Many studies of nanomaterials have shown limited GI absorption via the lymph duct, while others demonstrated low bioavailability of conventional drugs after oral administration (Yamago et al., 1995; Kreyling et al., 2002). Very low oral absorption of radiolabeled fullerenes and nanoparticles was previously observed in rats, and nanoparticle absorption via the GI tract was affected by particle size and surface properties (Jani et al., 1989; Hillyer and Albrecht, 2001). Small, hydrophobic, and neutral particles were more rapidly absorbed than large, hydrophilic, and charged particles (Hussain et al., 2001). Absorption of radiolabeled polystyrene nanoparticles was size dependent in rats (50 nm > 100 nm > 300–3000 nm) (Hillyer and Albrecht, 2001), with larger nanoparticles trapped in Peyer's patches (Jani et al., 1990).

4.8 Nanoscale World via the Buccal Route and Ingestion

The sizes of food contents, anatomical pores, and permeability barriers relevant to the buccal route and ingestion are described in this chapter and illustrated in Figure 4.14. The surface of the oral cavity, or the buccal mucosa, consists of nonkeratinous cells and has a paracellular permeability of 1 nm or less, whereas the pores in taste buds are ~20–30 μm in diameter. Type II gustatory cells in taste buds utilize the same signal transduction pathways as enteroneuroendocrine cells and translocate nanoparticles via a GPCR-dependent mechanism. Therefore, nanoparticles are more likely to be distributed by type II cells in the buccal route than by the nonkeratinized buccal mucosa. On the other hand, the small intestine exhibits the most permeable

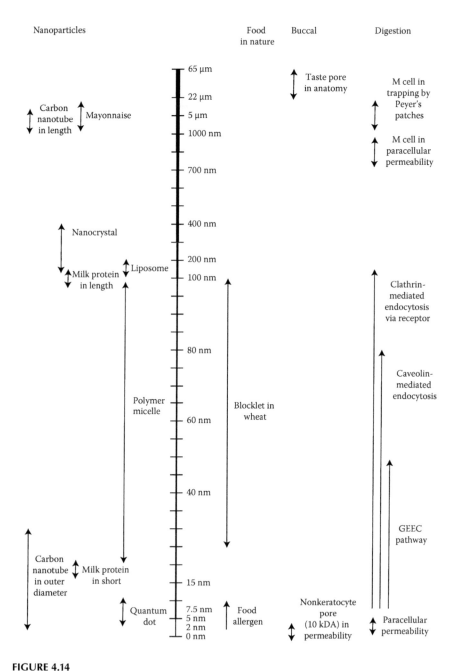

FIGURE 4.14
The size of nanoproducts administered via the buccal route and ingestion and the size of relevant anatomical features and permeability barriers.

mucosa in the GI tract because its large surface area contains numerous plicae circulares, villi, and microvilli.

A crystal size of 400 nm is recommended for practical applications of nanocrystal. In anatomy, the transcellular permeability cutoff size for absorptive cells is ~200 nm, given that the main clathrin-mediated pathway can only import substances of 200 nm or less. Paracellular transport of microparticles is primarily mediated by M cells. Nanoparticles of ≤200 nm can be endocytosed by M cells, and nanoparticles of ≤150–200 nm can be endocytosed by enteroneuroendocrine cells.

Finally, if we focus on the nanostructure of wheat as an example of a naturally occurring food, one starch granule is composed of a repeating block of 120 nm or less, matching the size of particles that are imported into cells by endocytosis. Food allergens tend to be 10 nm or less in size.

References

Aalberse, R. C. 1997. Food allergens. *Environmental Toxicology and Pharmacology* 4 (1–2):55–60. doi: 10.1016/S1382-6689(97)10042-4.

Abd El-Salam, M. H. and S. El-Shibiny. 2012. Formation and potential uses of milk proteins as nano delivery vehicles for nutraceuticals: A review. *International Journal of Dairy Technology* 65 (1):13–21. doi: 10.1111/j.1471-0307.2011.00737.x.

Ai, K. L., Y. L. Liu, and L. H. Lu. 2009. Hydrogen-bonding recognition-induced color change of gold nanoparticles for visual detection of melamine in raw milk and infant formula. *Journal of the American Chemical Society* 131 (27):9496–9497. doi: 10.1021/ja9037017.

Althues, H., J. Henle, and S. Kaskel. 2007. Functional inorganic nanofillers for transparent polymers. *Chemical Society Reviews* 36 (9):1454–1465. doi: 10.1039/b608177k.

Appendini, P. and J. H. Hotchkiss. 2002. Review of antimicrobial food packaging. *Innovative Food Science & Emerging Technologies* 3:113–126.

Arora, A. and G. W. Padua. 2010. Review: Nanocomposites in food packaging. *Journal of Food Science* 75 (1):R43–R49. doi: 10.1111/j.1750-3841.2009.01456.x.

Arvanitoyannis, I. S. and L. Bosnea. 2004. Migration of substances from food packaging materials to foods. *Critical Reviews in Food Science and Nutrition* 44 (2):63–76. doi: 10.1080/10408690490424621.

Averina, E. S. and I. A. Kutyrev. 2011. Perspectives on the use of marine and freshwater hydrobiont oils for development of drug delivery systems. *Biotechnology Advances* 29 (5):548–557. doi: 10.1016/j.biotechadv.2011.01.009.

Baker, C., A. Pradhan, L. Pakstis, D. J. Pochan, and S. I. Shah. 2005. Synthesis and antibacterial properties of silver nanoparticles. *Journal of Nanoscience and Nanotechnology* 5 (2):244–249. doi: 10.1166/jnn.2005.034.

Balogh, L., D. R. Swanson, D. A. Tomalia, G. L. Hagnauer, and A. T. McManus. 2001. Dendrimer-silver complexes and nanocomposites as antimicrobial agents. *Nano Letters* 1 (1):18–21. doi: 10.1021/nl005502p.

Barnett, M. L. and G. Szabo. 1973. Gap junctions in human gingival keratinized epithelium. *Journal of Periodontal Research* 8 (3):117–126.

Behrens, I., A. I. V. Pena, M. J. Alonso, and T. Kissel. 2002. Comparative uptake studies of bioadhesive and non-bioadhesive nanoparticles in human intestinal cell lines and rats: The effect of mucus on particle adsorption and transport. *Pharmaceutical Research* 19 (8):1185–1193.

Behrens, M., W. Meyerhof, C. Hellfritsch, and T. Hofmann. 2011. Sweet and umami taste: Natural products, their chemosensory targets, and beyond. *Angewandte Chemie—International Edition* 50 (10):2220–2242. doi: 10.1002/anie.201002094.

Benn, T., B. Cavanagh, K. Hristovski, J. D. Posner, and P. Westerhoff. 2010. The release of nanosilver from consumer products used in the home. *Journal of Environmental Quality* 39 (6):1875–1882. doi: 10.2134/jeq2009.0363.

Benn, T. M. and P. Westerhoff. 2008. Nanoparticle silver released into water from commercially available sock fabrics. *Environmental Science & Technology* 42 (11):4133–4139. doi: 10.1021/Es7032718.

Bernkop-Schnurch, A., C. E. Kast, and D. Guggi. 2003. Permeation enhancing polymers in oral delivery of hydrophilic macromolecules: Thiomer/GSH systems. *Journal of Controlled Release* 93 (2):95–103. doi: 10.1016/j.jconrel.2003.05.001.

Bhupathiraju, S. N. and K. L. Tucker. 2011. Coronary heart disease prevention: Nutrients, foods, and dietary patterns. *Clinica Chimica Acta* 412 (17–18): 1493–1514. doi: 10.1016/j.cca.2011.04.038.

Blanchette, J., N. Kavimandan, and N. A. Peppas. 2004. Principles of transmucosal delivery of therapeutic agents. *Biomedicine & Pharmacotherapy* 58 (3):142–151. doi: 10.1016/j.biopha.2004.01.006.

Borghesi, C., M. J. Taussig, and C. Nicoletti. 1999. Rapid appearance of M cells after microbial challenge is restricted at the periphery of the follicle-associated epithelium of Peyer's patch. *Laboratory Investigation* 79 (11):1393–1401.

Braun, T., P. Voland, L. Kunz, C. Prinz, and M. Gratzl. 2007. Enterochromaffin cells of the human gut: Sensors for spices and odorants. *Gastroenterology* 132 (5): 1890–1901. doi: 10.1053/j.gastro.2007.02.036.

Brayden, D. J. 2001. Oral vaccination in man using antigens in particles: Current status. *European Journal of Pharmaceutical Sciences* 14 (3):183–189. doi: 10.1016/s0928-0987(01)00175-0.

Brayden, D. J., M. A. Jepson, and A. W. Baird. 2005. Intestinal Peyer's patch M cells and oral vaccine targeting. *Drug Discovery Today* 10 (17):1145–1157. doi: 10.1016/s1359-6446(05)03536-1.

Braydich-Stolle, L., S. Hussain, J. J. Schlager, and M. C. Hofmann. 2005. In vitro cytotoxicity of nanoparticles in mammalian germline stem cells. *Toxicological Sciences* 88 (2):412–419. doi: 10.1093/toxsci/kfi256.

Canton, I. and G. Battaglia. 2012. Endocytosis at the nanoscale. *Chemical Society Reviews* 41 (7):2718–2739. doi: 10.1039/C2cs15309b.

Carino, G. P. and E. Mathiowitz. 1999. Oral insulin delivery. *Advanced Drug Delivery Reviews* 35 (2–3):249–257. doi: 10.1016/s0169-409x(98)00075-1.

Carlson, C., S. M. Hussain, A. M. Schrand, L. K. Braydich-Stolle, K. L. Hess, R. L. Jones, and J. J. Schlager. 2008. Unique cellular interaction of silver nanoparticles: Size-dependent generation of reactive oxygen species. *Journal of Physical Chemistry B* 112 (43):13608–13619. doi: 10.1021/Jp712087m.

Charlton, K. E., E. MacGregor, N. H. Vorster, N. S. Levitt, and K. Steyn. 2007. Partial replacement of NaCl can be achieved with potassium, magnesium and calcium salts in brown bread. *International Journal of Food Sciences and Nutrition* 58 (7):508–521. doi: 10.1080/09637480701331148.

Chaudhari, N. and S. D. Roper. 2010. The cell biology of taste. *Journal of Cell Biology* 190 (3):285–296. doi: 10.1083/jcb.201003144.

Chaudhry, Q., M. Scotter, J. Blackburn, B. Ross, A. Boxall, L. Castle, R. Aitken, and R. Watkins. 2008. Applications and implications of nanotechnologies for the food sector. *Food Additives and Contaminants* 25 (3):241–258. doi: 10.1080/02652030701744538.

Chawla, G., P. Gupta, V. Koradia, and A. K. Bansal. 2003. Gastroretention: A means to address regional variability in intestinal drug absorption. *Pharmaceutical Technology* 27:50–68.

Choi, S. J., E. A. Decker, L. Henson, L. M. Popplewell, and D. J. McClements. 2010. Inhibition of citral degradation in model beverage emulsions using micelles and reverse micelles. *Food Chemistry* 122 (1):111–116. doi: 10.1016/j.foodchem.2010.02.025.

Clark, M. A., M. A. Jepson, and B. H. Hirst. 2001. Exploiting M cells for drug and vaccine delivery. *Advanced Drug Delivery Reviews* 50 (1–2):81–106. doi: 10.1016/s0169-409x(01)00149-1.

Collins, L. M. C. and C. Dawes. 1987. The surface-area of the adult human mouth and thickness of the salivary film covering the teeth and oral-mucosa. *Journal of Dental Research* 66 (8):1300–1302.

Conner, S. D. and S. L. Schmid. 2003. Regulated portals of entry into the cell. *Nature* 422 (6927):37–44. doi: 10.1038/nature01451.

Corr, S. C., C. C. G. M. Gahan, and C. Hill. 2008. M-cells: Origin, morphology and role in mucosal immunity and microbial pathogenesis. *FEMS Immunology and Medical Microbiology* 52 (1):2–12. doi: 10.1111/j.1574-695X.2007.00359.x.

Cournarie, F., D. Auchere, D. Chevenne, B. Lacour, M. Seiller, and C. Vauthier. 2002. Absorption and efficiency of insulin after oral administration of insulin-loaded nanocapsules in diabetic rats. *International Journal of Pharmaceutics* 242 (1–2): 325–328. doi: 10.1016/s0378-5173(02)00175-8.

Damge, C., C. Michel, M. Aprahamian, and P. Couvreur. 1988. New approach for oral-administration of insulin with polyalkylcyanoacrylate nanocapsules as drug carrier. *Diabetes* 37 (2):246–251.

Damge, C., C. Michel, M. Aprahamian, P. Couvreur, and J. P. Devissaguet. 1990. Nanocapsules as carriers for oral peptide delivery. *Journal of Controlled Release* 13 (2–3):233–239. doi: 10.1016/0168-3659(90)90013-j.

des Rieux, A., V. Fievez, M. Garinot, Y. J. Schneider, and V. Preat. 2006. Nanoparticles as potential oral delivery systems of proteins and vaccines: A mechanistic approach. *Journal of Controlled Release* 116 (1):1–27. doi: 10.1016/j.jconrel.2006.08.013.

des Rieux, A., E. G. E. Ragnarsson, E. Gullberg, V. Preat, Y. J. Schneider, and P. Artursson. 2005. Transport of nanoparticles across an in vitro model of the human intestinal follicle associated epithelium. *European Journal of Pharmaceutical Sciences* 25 (4–5):455–465. doi: 10.1016/j.ejps.2005.04.015.

Deshpande, D. A., W. C. H. Wang, E. L. McIlmoyle, K. S. Robinett, R. M. Schillinger, S. S. An, J. S. K. Sham, and S. B. Liggett. 2010. Bitter taste receptors on airway smooth muscle bronchodilate by localized calcium signaling and reverse obstruction. *Nature Medicine* 16 (11):1299–1305. doi: 10.1038/nm.2237.

Dhaka, V., N. Gulia, K. S. Ahlawat, and B. S. Khatkar. 2011. Trans fats-sources, health risks and alternative approach—A review. *Journal of Food Science and Technology (Mysore)* 48 (5):534–541. doi: 10.1007/s13197-010-0225-8.

Diamond, G. L. and R. K. Zalups. 1998. Understanding renal toxicity of heavy metals. *Toxicologic Pathology* 26 (1):92–103.

Dickinson, E. 2011. Double emulsions stabilized by food biopolymers. *Food Biophysics* 6 (1):1–11. doi: 10.1007/s11483-010-9188-6.

Didierlaurent, A., J. C. Sirard, J. P. Kraehenbuhl, and M. R. Neutra. 2002. How the gut senses its content. *Cellular Microbiology* 4 (2):61–72.

Doherty, G. J. and H. T. McMahon. 2009. Mechanisms of endocytosis. *Annual Review of Biochemistry* 78:857–902. doi: 10.1146/annurev.biochem.78.081307.110540.

Duncan, T. V. 2011. Applications of nanotechnology in food packaging and food safety: Barrier materials, antimicrobials and sensors. *Journal of Colloid and Interface Science* 363 (1):1–24. doi: 10.1016/j.jcis.2011.07.017.

Eby, D. M., H. R. Luckarift, and G. R. Johnson. 2009. Hybrid antimicrobial enzyme and silver nanoparticle coatings for medical instruments. *ACS Applied Materials & Interfaces* 1 (7):1553–1560. doi: 10.1021/Am9002155.

Echague, C. G., P. S. Hair, and K. M. Cunnion. 2010. A comparison of antibacterial activity against methicillin-resistant *Staphylococcus aureus* and gram-negative organisms for antimicrobial compounds in a unique composite wound dressing. *Advances in Skin & Wound Care* 23 (9):406–413. doi: 10.1097/01. ASW.0000383213.95911.bc.

Eldridge, J. H., C. J. Hammond, J. A. Meulbroek, J. K. Staas, R. M. Gilley, and T. R. Tice. 1990. Controlled vaccine release in the gut-associated lymphoid-tissues. 1. Orally-administered biodegradable microspheres target the Peyer's patches. *Journal of Controlled Release* 11 (1–3):205–214. doi: 10.1016/0168-3659(90)90133-e.

Fasinu, P., V. Pillay, V. M. K. Ndesendo, L. C. du Toit, and Y. E. Choonara. 2011. Diverse approaches for the enhancement of oral drug bioavailability. *Biopharmaceutics & Drug Disposition* 32 (4):185–209. doi: 10.1002/Bdd.750.

Fathi, M., M. R. Mozafari, and M. Mohebbi. 2012. Nanoencapsulation of food ingredients using lipid based delivery systems. *Trends in Food Science & Technology* 23 (1):13–27. doi: 10.1016/j.tifs.2011.08.003.

FDA. 2009. Food additives permitted for direct addition to food for human consumption; silver nitrate and hydrogen peroxide. *Federal Register* 74 (51): 11476–11478.

Ferguson, M. L., K. Prasad, H. Boukari, D. L. Sackett, S. Krueger, E. M. Lafer, and R. Nossal. 2008. Clathrin triskelia show evidence of molecular flexibility. *Biophysical Journal* 95 (4):1945–1955. doi: 10.1529/biophysj.107.126342.

Florence, A. T. 2004. Issues in oral nanoparticle drug carrier uptake and targeting. *Journal of Drug Targeting* 12 (2):65–70. doi: 10.1080/10611860410001693706.

Florence, A. T. 2005. Nanoparticle uptake by the oral route: Fulfilling its potential? *Drug Delivery/Formulation and Nanotechnology* 2 (1):75–81. doi: 10.1080/ 10611860410001693706.

Foster, N. and B. H. Hirst. 2005. Exploiting receptor biology for oral vaccination with biodegradable particulates. *Advanced Drug Delivery Reviews* 57 (3):431–450. doi: 10.1016/j.addr.2004.09.009.

Fujita, H. and T. Fujita. 1992. *Textbook of Histology: Part 2*, 3rd edn. Igaku-Shoin Ltd., Tokyo, Japan.

Furno, F., K. S. Morley, B. Wong, B. L. Sharp, P. L. Arnold, S. M. Howdle, R. Bayston, P. D. Brown, P. D. Winship, and H. J. Reid. 2004. Silver nanoparticles and polymeric medical devices: A new approach to prevention of infection? *Journal of Antimicrobial Chemotherapy* 54 (6):1019–1024. doi: 10.1093/jac/dkh478.

Gabor, F., E. Bogner, A. Weissenboeck, and M. Wirth. 2004. The lectin-cell interaction and its implications to intestinal lectin-mediated drug delivery. *Advanced Drug Delivery Reviews* 56 (4):459–480. doi: 10.1016/j.addr.2003.10.015.

Gabor, F., M. Stangl, and M. Wirth. 1998. Lectin-mediated bioadhesion: Binding characteristics of plant lectins on the enterocyte-like cell lines Caco-2, HT-29 and HCT-8. *Journal of Controlled Release* 55 (2–3):131–142. doi: 10.1016/s0168-3659(98)00043-1.

Gallant, D. J., B. Bouchet, and P. M. Baldwin. 1997. Microscopy of starch: Evidence of a new level of granule organization. *Carbohydrate Polymers* 32 (3–4):177–191. doi: 10.1016/s0144-8617(97)00008-8.

Gebert, A., L. Steinmetz, S. Fassbender, and K. H. Wendlandt. 2004. Antigen transport into Peyer's patches—Increased uptake by constant numbers of M cells. *American Journal of Pathology* 164 (1):65–72. doi: 10.1016/s0002-9440(10)63097-0.

Gemmell, C. G., D. I. Edwards, A. P. Fraise, F. K. Gould, G. L. Ridgway, R. E. Warren, and Society Joint Working Party British. 2006. Guidelines for the prophylaxis and treatment of methicillin-resistant *Staphylococcus aureus* (MRSA) infections in the UK. *Journal of Antimicrobial Chemotherapy* 57 (4):589–608. doi: 10.1093/jac/dkl017.

Giannasca, P. J., K. T. Giannasca, A. M. Leichtner, and M. R. Neutra. 1999. Human intestinal M cells display the sialyl Lewis A antigen. *Infection and Immunity* 67 (2):946–953.

Graveland-Bikker, J. F. and C. G. de Kruif. 2006. Unique milk protein based nanotubes: Food and nanotechnology meet. *Trends in Food Science & Technology* 17 (5):196–203. doi: 10.1016/j.tifs.2005.12.009.

Graveland-Bikker, J. F., R. I. Koning, H. K. Koerten, R. B. J. Geels, R. M. A. Heeren, and C. G. de Kruif. 2009. Structural characterization of α-lactalbumin nanotubes. *Soft Matter* 5 (10):2020–2026. doi: 10.1039/B815775h.

Graveland-Bikker, J. F., I. A. T. Schaap, C. F. Schmidt, and C. G. de Kruif. 2006. Structural and mechanical study of a self-assembling protein nanotube. *Nano Letters* 6 (4):616–621. doi: 10.1021/Nl052205h.

Gref, R., P. Couvreur, G. Barratt, and E. Mysiakine. 2003. Surface-engineered nanoparticles for multiple ligand coupling. *Biomaterials* 24 (24):4529–4537. doi: 10.1016/s0142-9612(03)00348-x.

Gullberg, E., M. Leonard, J. Karlsson, A. M. Hopkins, D. Brayden, A. W. Baird, and P. Artursson. 2000. Expression of specific markers and particle transport in a new human intestinal M-cell model. *Biochemical and Biophysical Research Communications* 279 (3):808–813. doi: 10.1006/bbrc.2000.4038.

Harris, D. and J. R. Robinson. 1992. Drug delivery via the mucous-membranes of the oral cavity. *Journal of Pharmaceutical Sciences* 81 (1):1–10.

Heinemann, L. 2011. New ways of insulin delivery. *International Journal of Clinical Practice* 65:31–46. doi: 10.1111/j.1742-1241.2010.02577.x.

Helland, I. B., L. Smith, K. Saarem, O. D. Saugstad, and C. A. Drevon. 2003. Maternal supplementation with very-long-chain n-3 fatty acids during pregnancy and lactation augments children's IQ at 4 years of age. *Pediatrics* 111 (1):39–44.

Herrera, J. L., M. F. Lyons, and L. F. Johnson. 1988. Saliva—Its role in health and disease. *Journal of Clinical Gastroenterology* 10 (5):569–578.

Hillery, A. M. and A. T. Florence. 1996. The effect of adsorbed poloxamer 188 and 407 surfactants on the intestinal uptake of 60-nm polystyrene particles after oral administration in the rat. *International Journal of Pharmaceutics* 132 (1–2):123–130.

Hillyer, J. F. and R. M. Albrecht. 2001. Gastrointestinal persorption and tissue distribution of differently sized colloidal gold nanoparticles. *Journal of Pharmaceutical Sciences* 90 (12):1927–1936. doi: 10.1002/jps.1143.abs.

Hofman, P. 2003. Pathological interactions of bacteria and toxins with the gastrointestinal epithelial tight junctions and/or the zonula adherens: An update. *Cellular and Molecular Biology* 49 (1):65–75.

Hoogstraate, J. A. J. and P. W. Wertz. 1998. Drug delivery via the buccal mucosa. *Pharmaceutical Science & Technology Today* 1 (7):309–316.

Hunter, J. E., J. Zhang, and P. M. Kris-Etherton. 2010. Cardiovascular disease risk of dietary stearic acid compared with trans, other saturated, and unsaturated fatty acids: A systematic review. *American Journal of Clinical Nutrition* 91 (1):46–63. doi: 10.3945/ajcn.2009.27661.

Hussain, N. 2000. Ligand-mediated tissue specific drug delivery. *Advanced Drug Delivery Reviews* 43 (2–3):95–100. doi: 10.1016/s0169-409x(00)00066-1.

Hussain, N., V. Jaitley, and A. T. Florence. 2001. Recent advances in the understanding of uptake of microparticulates across the gastrointestinal lymphatics. *Advanced Drug Delivery Reviews* 50 (1–2):107–142. doi: 10.1016/s0169-409x(01)00152-1.

Igarashi, E. 2008. Factors affecting toxicity and efficacy of polymeric nanomedicines. *Toxicology and Applied Pharmacology* 229 (1):121–134. doi: 10.1016/j.taap.2008.02.007.

Ikeda, K. 2002. New seasonings. *Chemical Senses* 27 (9):847–849. doi: 10.1093/chemse/27.9.847.

Inoue, M. and S. Tsugane. 2005. Epidemiology of gastric cancer in Japan. *Postgraduate Medical Journal* 81 (957):419–424. doi: 10.1136/pgmj.2004.029330.

Iwase, Y. and Y. Maitani. 2011. Octreotide-targeted liposomes loaded with CPT-11 enhanced cytotoxicity for the treatment of medullary thyroid carcinoma. *Molecular Pharmaceutics* 8 (2):330–337. doi: 10.1021/Mp100380y.

Iwatsuki, K., R. Ichikawa, A. Uematsu, A. Kitamura, H. Uneyama, and K. Torii. 2012. Detecting sweet and umami tastes in the gastrointestinal tract. *Acta Physiologica* 204 (2):169–177. doi: 10.1111/j.1748-1716.2011.02353.x.

Jani, P., G. W. Halbert, J. Langridge, and A. T. Florence. 1989. The uptake and translocation of latex nanospheres and microspheres after oral-administration to rats. *Journal of Pharmacy and Pharmacology* 41 (12):809–812.

Jani, P., G. W. Halbert, J. Langridge, and A. T. Florence. 1990. Nanoparticle uptake by the rat gastrointestinal mucosa—Quantitation and particle-size dependency. *Journal of Pharmacy and Pharmacology* 42 (12):821–826.

Jeong, S. H., Y. H. Hwang, and S. C. Yi. 2005a. Antibacterial properties of padded PP/PE nonwovens incorporating nano-sized silver colloids. *Journal of Materials Science* 40 (20):5413–5418. doi: 10.1007/s10853-005-4340-2.

Jeong, S. H., S. Y. Yeo, and S. C. Yi. 2005b. The effect of filler particle size on the antibacterial properties of compounded polymer/silver fibers. *Journal of Materials Science* 40 (20):5407–5411. doi: 10.1007/s10853-005-4339-8.

Jepson, M. A., M. A. Clark, N. Foster, C. M. Mason, M. K. Bennett, N. L. Simmons, and B. H. Hirst. 1996. Targeting to intestinal M cells. *Journal of Anatomy* 189:507–516.

Jones, O. G., U. Lesmes, P. Dubin, and D. J. McClements. 2010. Effect of polysaccharide charge on formation and properties of biopolymer nanoparticles created by heat treatment of β-lactoglobulin-pectin complexes. *Food Hydrocolloids* 24 (4): 374–383. doi: 10.1016/j.foodhyd.2009.11.003.

Jones, S. A., P. G. Bowler, M. Walker, and D. Parsons. 2004. Controlling wound bio-burden with a novel silver-containing hydrofiber(R) dressing. *Wound Repair and Regeneration* 12 (3):288–294. doi: 10.1111/j.1067-1927.2004.012304.x.

Jung, T., W. Kamm, A. Breitenbach, E. Kaiserling, J. X. Xiao, and T. Kissel. 2000. Biodegradable nanoparticles for oral delivery of peptides: Is there a role for poly-mers to affect mucosal uptake? *European Journal of Pharmaceutics and Biopharma-ceutics* 50 (1):147–160. doi: 10.1016/s0939-6411(00)00084-9.

Junginger, H. E., J. A. Hoogstraate, and J. C. Verhoef. 1999. Recent advances in buccal drug delivery and absorption—In vitro and in vivo studies. *Journal of Controlled Release* 62 (1–2):149–159.

Kalra, S., B. Kalra, and N. Agrawal. 2010. Oral insulin. *Diabetology & Metabolic Syndrome* 2:66. doi: 10.1186/1758-5996-2-66.

Kampmann, Y., E. De Clerck, S. Kohn, D. K. Patchala, R. Langerock, and J. Kreyenschmidt. 2008. Study on the antimicrobial effect of silver-containing inner liners in refrigerators. *Journal of Applied Microbiology* 104 (6):1808–1814. doi: 10.1111/j.1365-2672.2008.03727.x.

Kararli, T. T. 1995. Comparison of the gastrointestinal anatomy, physiology, and bio-chemistry of humans and commonly used laboratory-animals. *Biopharmaceutics & Drug Disposition* 16 (5):351–380.

Kawata, K., M. Osawa, and S. Okabe. 2009. In vitro toxicity of silver nanoparticles at noncytotoxic doses to HepG2 human hepatoma cells. *Environmental Science & Technology* 43 (15):6046–6051. doi: 10.1021/es900754q.

Kelly, G. 2010. A review of the sirtuin system, its clinical implications, and the poten-tial role of dietary activators like resveratrol: Part 1. *Alternative Medicine Review* 15 (3):245–263.

Kettiger, H., A. Schipanski, P. Wick, and J. Huwyler. 2013. Engineered nanomaterial uptake and tissue distribution: From cell to organism. *International Journal of Nanomedicine* 8:3255–3269. doi: 10.2147/Ijn.S49770.

Khaksar, M. R. and M. Ghazi-Khansari. 2009. Determination of migration monomer styrene from GPPS (general purpose polystyrene) and HIPS (high impact poly-styrene) cups to hot drinks. *Toxicology Mechanisms and Methods* 19 (3):257–261. doi: 10.1080/15376510802510299.

Kim, J. S., E. Kuk, K. N. Yu, J. H. Kim, S. J. Park, H. J. Lee, S. H. Kim et al. 2007. Antimicrobial effects of silver nanoparticles. *Nanomedicine—Nanotechnology Biology and Medicine* 3 (1):95–101. doi: 10.1016/j.nano.2006.12.001.

Kinnamon, S. C. 2012. Taste receptor signalling—From tongues to lungs. *Acta Physiologica* 204 (2):158–168. doi: 10.1111/j.1748-1716.2011.02308.x.

Klaypradit, W. and Y. W. Huang. 2008. Fish oil encapsulation with chitosan using ultrasonic atomizer. *LWT—Food Science and Technology* 41 (6):1133–1139. doi: 10.1016/j.lwt.2007.06.014.

Kounosu, M. and S. Kaneko. 2007. Antibacterial activity of antibacterial cutting boards in household kitchens. *Biocontrol Science* 12 (4):123–130.

Kraehenbuhl, J. P. and M. R. Neutra. 1992. Molecular and cellular basis of immune protection of mucosal surfaces. *Physiological Reviews* 72 (4):853–879.

Kreyling, W. G., M. Semmler, F. Erbe, P. Mayer, S. Takenaka, H. Schulz, G. Oberdorster, and A. Ziesenis. 2002. Translocation of ultrafine insoluble iridium particles from lung epithelium to extrapulmonary organs is size dependent but very low. *Journal of Toxicology and Environmental Health—Part A* 65 (20):1513–1530. doi: 10.1080/00984100290071649.

Kris-Etherton, P. M. 2010. Trans-fats and coronary heart disease. *Critical Reviews in Food Science and Nutrition* 50:29–30. pii: 930706839, doi: 10.1080/10408398. 2010.526872.

Kroese, F. M., C. Evers, and D. T. D. De Ridder. 2009. How chocolate keeps you slim. The effect of food temptations on weight watching goal importance, intentions, and eating behavior. *Appetite* 53 (3):430–433. doi: 10.1016/j.appet.2009.08.002.

Kuhlman, J. J., S. Lalani, J. Magluilo, B. Levine, W. D. Darwin, R. E. Johnson, and E. J. Cone. 1996. Human pharmacokinetics of intravenous, sublingual, and buccal buprenorphine. *Journal of Analytical Toxicology* 20 (6):369–378.

Kurosaki, Y. and T. Kimura. 2000. Regional variation in oral mucosal drug permeability. *Critical Reviews in Therapeutic Drug Carrier Systems* 17 (5):467–508.

Kvitek, L., A. Panacek, J. Soukupova, M. Kolar, R. Vecerova, R. Prucek, M. Holecova, and R. Zboril. 2008. Effect of surfactants and polymers on stability and antibacterial activity of silver nanoparticles (NPs). *Journal of Physical Chemistry C* 112 (15):5825–5834. doi: 10.1021/jp711616v.

Laca, A., B. Paredes, and M. Diaz. 2012. Lipid-enriched egg yolk fraction as ingredient in cosmetic emulsions. *Journal of Texture Studies* 43 (1):12–28. doi: 10.1111/j.1745-4603.2011.00312.x.

Laca, A., M. C. Saenz, B. Paredes, and M. Diaz. 2010. Rheological properties, stability and sensory evaluation of low-cholesterol mayonnaises prepared using egg yolk granules as emulsifying agent. *Journal of Food Engineering* 97 (2):243–252. doi: 10.1016/j.jfoodeng.2009.10.017.

Lamprecht, A., U. Schafer, and C. M. Lehr. 2001. Size-dependent bioadhesion of micro- and nanoparticulate carriers to the inflamed colonic mucosa. *Pharmaceutical Research* 18 (6):788–793. doi: 10.1023/a:1011032328064.

Lavker, R. M. 1976. Membrane coating granules—Fate of discharged lamellae. *Journal of Ultrastructure Research* 55 (1):79–86.

LeFevre, M. E., J. B. Warren, and D. D. Joel. 1985. Particles and macrophages in murine Peyer's patches. *Experimental Cell Biology* 53 (3):121–129.

Lehr, C. M., F. G. J. Poelma, H. E. Junginger, and J. J. Tukker. 1991. An estimate of turnover time of intestinal mucus gel layer in the rat in situ loop. *International Journal of Pharmaceutics* 70 (3):235–240. doi: 10.1016/0378-5173(91)90287-x.

Lesch, C. A., C. A. Squier, A. Cruchley, D. M. Williams, and P. Speight. 1989. The permeability of human oral-mucosa and skin to water. *Journal of Dental Research* 68 (9):1345–1349.

Ley, R. E., D. A. Peterson, and J. I. Gordon. 2006. Ecological and evolutionary forces shaping microbial diversity in the human intestine. *Cell* 124 (4):837–848. doi: 10.1016/j.cell.2006.02.017.

List, M. and H. Sucker. 1988. Pharmaceutical colloidal hydrosols for injection. GB patent: 2200048.

Liversidge, G. G., K. C. Cundy, J. F. Bishop, and D. A. Czekai. 1992. Surface modified drug nanoparticles. U.S. patent: 5145684.

Livney, Y. D. 2010. Milk proteins as vehicles for bioactives. *Current Opinion in Colloid & Interface Science* 15 (1–2):73–83. doi: 10.1016/j.cocis.2009.11.002.

Lochner, N., F. Pittner, M. Wirth, and F. Gabor. 2003. Wheat germ agglutinin binds to the epidermal growth factor receptor of artificial Caco-2 membranes as detected by silver nanoparticle enhanced fluorescence. *Pharmaceutical Research* 20 (5): 833–839. doi: 10.1023/a:1023406224028.

Lok, C. N., C. M. Ho, R. Chen, Q. Y. He, W. Y. Yu, H. Sun, P. K. H. Tam, J. F. Chiu, and C. M. Che. 2007. Silver nanoparticles: Partial oxidation and antibacterial activities. *Journal of Biological Inorganic Chemistry* 12 (4):527–534. doi: 10.1007/s00775-007-0208-z.

Lundqvist-Persson, C., G. Lau, P. Nordin, B. Strandvik, and K. G. Sabel. 2010. Early behaviour and development in breast-fed premature infants are influenced by ω-6 and ω-3 fatty acid status. *Early Human Development* 86 (7):407–412. doi: 10.1016/j.earlhumdev.2010.05.017.

Luther, E. M., Y. Koehler, J. Diendorf, M. Epple, and R. Dringen. 2011. Accumulation of silver nanoparticles by cultured primary brain astrocytes. *Nanotechnology* 22 (37):375101. doi: 10.1088/0957-4484/22/37/375101.

Ma, Z. S., T. M. Lim, and L. Y. Lim. 2005. Pharmacological activity of peroral chitosan-insulin nanoparticles in diabetic rats. *International Journal of Pharmaceutics* 293 (1–2):271–280. doi: 10.1016/j.ijpharm.2004.12.025.

Mantis, N. J., M. C. Cheung, K. R. Chintalacharuvu, J. Rey, B. Corthesy, and M. R. Neutra. 2002. Selective adherence of IgA to murine Peyer's patch M cells: Evidence for a novel IgA receptor. *Journal of Immunology* 169 (4):1844–1851.

Martinez-Castanon, G. A., N. Nino-Martinez, F. Martinez-Gutierrez, J. R. Martinez-Mendoza, and F. Ruiz. 2008. Synthesis and antibacterial activity of silver nanoparticles with different sizes. *Journal of Nanoparticle Research* 10 (8): 1343–1348. doi: 10.1007/s11051-008-9428-6.

McClements, D. J. 2010. Emulsion design to improve the delivery of functional lipophilic components. *Annual Review of Food Science and Technology*, 1 (1):241–269. doi: 10.1146/annurev.food.080708.100722.

McNamara, R. K. and S. E. Carlson. 2006. Role of ω-3 fatty acids in brain development and function: Potential implications for the pathogenesis and prevention of psychopathology. *Prostaglandins Leukotrienes and Essential Fatty Acids* 75 (4–5): 329–349. doi: 10.1016/j.plefa.2006.07.010.

Mescher, A. L. 2010. *Junqueira's Basic Histology*, 12th edn. McGraw-Hill, New York.

Meyer, T., D. Akimov, N. Tarcea, S. Chatzipapadopoulos, G. Muschiolik, J. Kobow, M. Schmitt, and J. Popp. 2008. Three-dimensional molecular mapping of a multiple emulsion by means of CARS microscopy. *Journal of Physical Chemistry B* 112 (5):1420–1426. doi: 10.1021/Jp709643h.

Meyerhof, W., C. Batram, C. Kuhn, A. Brockhoff, E. Chudoba, B. Bufe, G. Appendino, and M. Behrens. 2010. The molecular receptive ranges of human TAS2R bitter taste receptors. *Chemical Senses* 35 (2):157–170. doi: 10.1093/chemse/bjp092.

Meynell, H. M., N. W. Thomas, P. S. James, J. Holland, M. J. Taussig, and C. Nicoletti. 1999. Up-regulation of microsphere transport across the follicle-associated epithelium of Peyer's patch by exposure to *Streptococcus pneumoniae* R36a. *FASEB Journal* 13 (6):611–619.

Mikos, A. G., E. Mathiowitz, R. Langer, and N. A. Peppas. 1991. Interaction of polymer microsphere with mucin gels as a means of characterizing polymer retention on mucus. *Journal of Colloid and Interface Science* 143 (2):366–373. doi: 10.1016/0021-9797(91)90270-i.

Miller, I. J. 1989. Variation in human taste bud density as a function of age. *Annals of the New York Academy of Sciences* 561:307–319. doi: 10.1111/j.1749-6632.1989.tb20991.x.

Mills, A. 2005. Oxygen indicators and intelligent inks for packaging food. *Chemical Society Reviews* 34 (12):1003–1011. doi: 10.1039/b503997p.

Mohan, S. and N. R. C. Campbell. 2009. Salt and high blood pressure. *Clinical Science* 117 (1–2):1–11. doi: 10.1042/Cs20080207.

Morones, J. R., J. L. Elechiguerra, A. Camacho, K. Holt, J. B. Kouri, J. T. Ramirez, and M. J. Yacaman. 2005. The bactericidal effect of silver nanoparticles. *Nanotechnology* 16 (10):2346–2353. doi: 10.1088/0957-4484/16/10/059.

Morris, M. C., D. A. Evans, C. C. Tangney, J. L. Bienias, and R. S. Wilson. 2005. Fish consumption and cognitive decline with age in a large community study. *Archives of Neurology* 62 (12):1849–1853.

Moss, J. 2006. Labeling of trans fatty acid content in food, regulations and limits—The FDA view. *Atherosclerosis Supplements* 7 (2):57–59. doi: 10.1016/j.atherosclerosissup.2006.04.012.

Mozafari, M. R., J. Flanagan, L. Matia-Merino, A. Awati, A. Omri, Z. E. Suntres, and H. Singh. 2006. Recent trends in the lipid-based nanoencapsulation of antioxidants and their role in foods. *Journal of the Science of Food and Agriculture* 86 (13):2038–2045. doi: 10.1002/Jsfa.2576.

Muncke, J. 2009. Exposure to endocrine disrupting compounds via the food chain: Is packaging a relevant source? *Science of the Total Environment* 407 (16):4549–4559. doi: 10.1016/j.scitotenv.2009.05.006.

Nellans, H. N. 1991. Mechanisms of peptide and protein-absorption. 1. Paracellular intestinal transport-modulation of absorption. *Advanced Drug Delivery Reviews* 7 (3):339–364. doi: 10.1016/0169-409x(91)90013-3.

Neutra, M. R., T. L. Phillips, E. L. Mayer, and D. J. Fishkind. 1987. Transport of membrane-bound macromolecules by M-cells in follicle-associated epithelium of rabbit Peyer's patch. *Cell and Tissue Research* 247 (3):537–546. doi: 10.1007/bf00215747.

Nicolazzo, J. A., B. L. Reed, and B. C. Finnin. 2005. Buccal penetration enhancers—How do they really work? *Journal of Controlled Release* 105 (1–2):1–15. doi: 10.1016/j.jconrel.2005.01.024.

Norris, D. A. and P. J. Sinko. 1997. Effect of size, surface charge, and hydrophobicity on the translocation of polystyrene microspheres through gastrointestinal mucin. *Journal of Applied Polymer Science* 63 (11):1481–1492.

Nowack, B., H. F. Krug, and M. Height. 2011. 120 Years of nanosilver history: Implications for policy makers. *Environmental Science & Technology* 45 (4): 1177–1183. doi: 10.1021/Es103316q.

Ohashi, K. and A. Shimada. 2005. Effect of salt type on the emulsifying properties of simulated-mayonnaise oil-in-water emulsions prepared with diacylglycerol. *Journal of the Japanese Society for Food Science and Technology–Nippon Shokuhin Kagaku Kogaku Kaishi* 52 (5):226–235.

Ohsu, T., Y. Amino, H. Nagasaki, T. Yamanaka, S. Takeshita, T. Hatanaka, Y. Maruyama, N. Miyamura, and Y. Eto. 2010. Involvement of the calcium-sensing receptor in human taste perception. *Journal of Biological Chemistry* 285 (2):1016–1022. doi: 10.1074/jbc.M109.029165.

Owen, R. L. 1977. Sequential uptake of horseradish-peroxidase by lymphoid follicle epithelium of Peyer's patches in normal unobstructed mouse intestine-ultrastructural-study. *Gastroenterology* 72 (3):440–451.

Pal, S., Y. K. Tak, and J. M. Song. 2007. Does the antibacterial activity of silver nanoparticles depend on the shape of the nanoparticle? A study of the gram-negative bacterium *Escherichia coli*. *Applied and Environmental Microbiology* 73 (6): 1712–1720. doi: 10.1128/aem.02218-06.

Panacek, A., L. Kvitek, R. Prucek, M. Kolar, R. Vecerova, N. Pizurova, V. K. Sharma, T. Nevecna, and R. Zboril. 2006. Silver colloid nanoparticles: Synthesis, characterization, and their antibacterial activity. *Journal of Physical Chemistry B* 110 (33):16248–16253. doi: 10.1021/jp063826h.

Pang, K. S. 2003. Modeling of intestinal drug absorption: Roles of transporters and metabolic enzymes (for the Gillette Review Series). *Drug Metabolism and Disposition* 31 (12):1507–1519.

Panyala, N. R., E. M. Pena-Mendez, and J. Havel. 2008. Silver or silver nanoparticles: A hazardous threat to the environment and human health? *Journal of Applied Biomedicine* 6 (3):117–129.

Pather, S. I., M. J. Rathbone, and S. Senel. 2008. Current status and the future of buccal drug delivery systems. *Expert Opinion on Drug Delivery* 5 (5):531–542. doi: 10.1517/17425240802085633.

Peppas, N. A. and D. A. Carr. 2009. Impact of absorption and transport on intelligent therapeutics and nanoscale delivery of protein therapeutic agents. *Chemical Engineering Science* 64 (22):4553–4565. doi: 10.1016/j.ces.2009.04.050.

Perakslis, E., A. Tuesca, and A. Lowman. 2007. Complexation hydrogels for oral protein delivery: An in vitro assessment of the insulin transport-enhancing effects following dissolution in simulated digestive fluids. *Journal of Biomaterials Science—Polymer Edition* 18 (12):1475–1490.

Pinto-Alphandary, H., M. Aboubakar, D. Jaillard, P. Couvreur, and C. Vauthier. 2003. Visualization of insulin-loaded nanocapsules: In vitro and in vivo studies after oral administration to rats. *Pharmaceutical Research* 20 (7):1071–1084.

Pocas, M. D. and T. Hogg. 2007. Exposure assessment of chemicals from packaging materials in foods: A review. *Trends in Food Science & Technology* 18 (4):219–230. doi: 10.1016/j.tifs.2006.008.

Porta, C., P. S. James, A. D. Phillips, T. C. Savidge, M. W. Smith, and D. Cremaschi. 1992. Confocal analysis of fluorescent bead uptake by mouse Peyer's patch follicle-associated M-cells. *Experimental Physiology* 77 (6):929–932.

Powers, C. M., N. Wrench, I. T. Ryde, A. M. Smith, F. J. Seidler, and T. A. Slotkin. 2010. Silver impairs neurodevelopment: Studies in PC12 cells. *Environmental Health Perspectives* 118 (1):73–79. doi: 10.1289/Ehp.0901149.

Pusztai, A., S. W. B. Ewen, G. Grant, D. S. Brown, J. C. Stewart, W. J. Peumans, E. J. M. Vandamme, and S. Bardocz. 1993. Antinutritive effects of wheat-germagglutinin and other N-acetylglucosamine-specific lectins. *British Journal of Nutrition* 70 (1):313–321. doi: 10.1079/bjn19930124.

Quintavalla, S. and L. Vicini. 2002. Antimicrobial food packaging in meat industry. *Meat Science* 62 (3):373–380.

Rassing, M. R. 1994. Chewing gum as a drug-delivery system. *Advanced Drug Delivery Reviews* 13 (1–2):89–121.

Rejman, J., V. Oberle, I. S. Zuhorn, and D. Hoekstra. 2004. Size-dependent internalization of particles via the pathways of clathrin- and caveolae-mediated endocytosis. *Biochemical Journal* 377:159–169. doi: 10.1042/Bj20031253.

Rimoldi, M. and M. Rescig. 2005. Uptake and presentation of orally administered antigens. *Vaccine* 23 (15):1793–1796. doi: 10.1016/j.vaccine.2004.11.007.

Roe, D., B. Karandikar, N. Bonn-Savage, B. Gibbins, and J. B. Roullet. 2008. Antimicrobial surface functionalization of plastic catheters by silver nanoparticles. *Journal of Antimicrobial Chemotherapy* 61 (4):869–876. doi: 10.1093/jac/dkn034.

Romanowski, A. W., C. A. Squier, and C. A. Lesch. 1988. Permeability of rodent junctional epithelium to exogenous protein. *Journal of Periodontal Research* 23 (2):81–86.

Ross, M. H. and W. Pawlina. 2011. *Histology: A Text and Atlas*, 6th edn. Baltimore, MD: Lippincott Williams & Wilkins.

Rouge, N., P. Buri, and E. Doelker. 1996. Drug absorption sites in the gastrointestinal tract and dosage forms for site-specific delivery. *International Journal of Pharmaceutics* 136 (1–2):117–139.

Roy, M. J. and M. Varvayanis. 1987. Development of dome epithelium in gut-associated lymphoid-tissues-association of IgA with M-cells. *Cell and Tissue Research* 248 (3):645–651. doi: 10.1007/bf00216495.

Russell-Jones, G. J., H. Veitch, and L. Arthur. 1999. Lectin-mediated transport of nanoparticles across Caco-2 and OK cells. *International Journal of Pharmaceutics* 190 (2):165–174. doi: 10.1016/s0378-5173(99)00254-9.

Salman, H. H., C. Gamazo, M. A. Campanero, and J. M. Irache. 2005. Salmonella-like bioadhesive nanoparticles. *Journal of Controlled Release* 106 (1–2):1–13. doi: 10.1016/j.jconrel.2005.03.033.

Sarkar, S., A. D. Jana, S. K. Samanta, and G. Mostafa. 2007. Facile synthesis of silver nano particles with highly efficient anti-microbial property. *Polyhedron* 26 (15):4419–4426. doi: 10.1016/j.poly.2007.05.056.

Schaff, E. A., R. DiCenzo, and S. L. Fielding. 2005. Comparison of misoprostol plasma concentrations following buccal and sublingual administration. *Contraception* 71 (1):22–25. doi: 10.1016/j.contraception.2004.06.014.

Segovia, C., I. Hutchinson, D. G. Laing, and A. L. Jinks. 2002. A quantitative study of fungiform papillae and taste pore density in adults and children. *Developmental Brain Research* 138 (2):135–146.

Shakweh, M., M. Besnard, V. Nicolas, and E. Fattal. 2005. Poly (lactide-co-glycolide) particles of different physicochemical properties and their uptake by Peyer's patches in mice. *European Journal of Pharmaceutics and Biopharmaceutics* 61 (1–2): 1–13. doi: 10.1016/j.ejpb.2005.04.006.

Sharma, R., E. J. M. vanDamme, W. J. Peumans, P. Sarsfield, and U. Schumacher. 1996. Lectin binding reveals divergent carbohydrate expression in human and mouse Peyer's patches. *Histochemistry and Cell Biology* 105 (6):459–465. doi: 10.1007/bf01457659.

Sharp, D. 2004. Labelling salt in food: If yes, how? *Lancet* 364 (9451):2079–2081.

Shegokar, R. and R. H. Muller. 2010. Nanocrystals: Industrially feasible multifunctional formulation technology for poorly soluble actives. *International Journal of Pharmaceutics* 399 (1–2):129–139. doi: 10.1016/j.ijpharm.2010.07.044.

Silva, A. S., J. M. C. Freire, R. Sendon, R. Franz, and P. P. Losada. 2009. Migration and diffusion of diphenylbutadiene from packages into foods. *Journal of Agricultural and Food Chemistry* 57 (21):10225–10230. doi: 10.1021/jf901666h.

Simopoulos, A. P., A. Leaf, and N. Salem. 2000. Workshop statement on the essentiality of and recommended dietary intakes for ω-6 and ω-3 fatty acids. *Prostaglandins Leukotrienes and Essential Fatty Acids* 63 (3):119–121. doi: 10.1054/plef.2000.0176.

Smart, J. D. 2005. The basics and underlying mechanisms of mucoadhesion. *Advanced Drug Delivery Reviews* 57 (11):1556–1568. doi: 10.1016/j.addr.2005.07.001.

Smith, B. K., L. E. Robinson, R. Nam, and D. W. L. Ma. 2009. Trans-fatty acids and cancer: A mini-review. *British Journal of Nutrition* 102 (9):1254–1266. doi: 10.1017/S0007114509991437.

Smith, J. M., M. Dornish, and E. J. Wood. 2005. Involvement of protein kinase C in chitosan glutamate-mediated tight junction disruption. *Biomaterials* 26 (16): 3269–3276. doi: 10.1016/j.biomaterials.2004.06.020.

Smith, P. J., M. Giroud, H. L. Wiggins, F. Gower, J. A. Thorley, B. Stolpe, J. Mazzolini, R. J. Dyson, and J. Z. Rappoport. 2012. Cellular entry of nanoparticles via serum sensitive clathrin-mediated endocytosis, and plasma membrane permeabilization. *International Journal of Nanomedicine* 7:2045–2055. doi: 10.2147/Ijn.S29334.

Sohi, H., A. Ahuja, F. J. Ahmad, and R. K. Khar. 2010. Critical evaluation of permeation enhancers for oral mucosal drug delivery. *Drug Development and Industrial Pharmacy* 36 (3):254–282. doi: 10.3109/03639040903117348.

Sondi, I. and B. Salopek-Sondi. 2004. Silver nanoparticles as antimicrobial agent: A case study on *E. coli* as a model for Gram-negative bacteria. *Journal of Colloid and Interface Science* 275 (1):177–182. doi: 10.1016/j.jcis.2004.02.012.

Song, Y., Y. Wang, R. Thakur, V. M. Meidan, and B. Michniak. 2004. Mucosal drug delivery: Membranes, methodologies, and applications. *Critical Reviews in Therapeutic Drug Carrier Systems* 21 (3):195–256.

Squier, C. A., P. Cox, and P. W. Wertz. 1991. Lipid-content and water permeability of skin and oral-mucosa. *Journal of Investigative Dermatology* 96 (1):123–126.

Squier, C. A. and M. J. Kremer. 2001. Biology of oral mucosa and esophagus. *Journal of the National Cancer Institute Monographs* 29:7–15.

Squier, C. A. and D. Nanny. 1985. Measurement of blood-flow in the oral-mucosa and skin of the rhesus-monkey using radiolabeled microspheres. *Archives of Oral Biology* 30 (4):313–318.

Squier, C. A. and P. W. Wertz. 1993. Permeability and the pathophysiology of oral-mucosa. *Advanced Drug Delivery Reviews* 12 (1–2):13–24.

Srur, E., O. Stachs, R. Guthoff, M. Witt, H. W. Pau, and T. Just. 2010. Change of the human taste bud volume over time. *Auris Nasus Larynx* 37 (4):449–455. doi: 10.1016/j.anl.2009.11.010.

Stender, S. and J. Dyerberg. 2004. Influence of trans fatty acids on health. *Annals of Nutrition and Metabolism* 48 (2):61–66. doi: 10.1159/000075591.

Stephen, A. M. and J. H. Cummings. 1980. Microbial contribution to human fecal mass. *Journal of Medical Microbiology* 13 (1):45–56.

Sucker, H. and P. Gassmann. 1994. Improvements in pharmaceutical compositions. GB patent: 2269536A.

Tada, N., C. Maruyama, S. Koba, H. Tanaka, S. Birou, T. Teramoto, and J. Sasaki. 2011. Japanese dietary lifestyle and cardiovascular disease. *Journal of Atherosclerosis and Thrombosis* 18 (9):723–734.

Tanaka, T. 1984. Transport pathway and uptake of microperoxidase in the junctional epithelium of healthy rat gingiva. *Journal of Periodontal Research* 19 (1):26–39.

Tang, J. L., L. Xiong, S. Wang, J. Y. Wang, L. Liu, J. G. Li, F. Q. Yuan, and T. F. Xi. 2009. Distribution, translocation and accumulation of silver nanoparticles in rats. *Journal of Nanoscience and Nanotechnology* 9 (8):4924–4932. doi: 10.1166/jnn.2009.1269.

Tizzano, M., M. Cristofoletti, A. Sbarbati, and T. E. Finger. 2011. Expression of taste receptors in solitary chemosensory cells of rodent airways. *BMC Pulmonary Medicine* 11:3. doi: 10.1186/1471-2466-11-3.

Tizzano, M., B. D. Gulbransen, A. Vandenbeuch, T. R. Clapp, J. P. Herman, H. M. Sibhatu, M. E. A. Churchill, W. L. Silver, S. C. Kinnamon, and T. E. Finger. 2010. Nasal chemosensory cells use bitter taste signaling to detect irritants and bacterial signals. *Proceedings of the National Academy of Sciences of the United States of America* 107 (7):3210–3215. doi: 10.1073/pnas.0911934107.

Tobio, M., R. Gref, A. Sanchez, R. Langer, and M. J. Alonso. 1998. Stealth PLA-PEG nanoparticles as protein carriers for nasal administration. *Pharmaceutical Research* 15 (2):270–275. doi: 10.1023/a:1011922819926.

Tobio, M., A. Sanchez, A. Vila, I. Soriano, C. Evora, J. L. Vila-Jato, and M. J. Alonso. 2000. The role of PEG on the stability in digestive fluids and in vivo fate of PEG-PLA nanoparticles following oral administration. *Colloids and Surfaces B— Biointerfaces* 18 (3–4):315–323. doi: 10.1016/s0927-7765(99)00157-5.

Vila, A., H. Gill, O. McCallion, and M. J. Alonso. 2004. Transport of PLA-PEG particles across the nasal mucosa: Effect of particle size and PEG coating density. *Journal of Controlled Release* 98 (2):231–244. doi: 10.1016/j.jconrel.2004.04.026.

Vila, A., A. Sanchez, M. Tobio, P. Calvo, and M. J. Alonso. 2002. Design of biodegradable particles for protein delivery. *Journal of Controlled Release* 78 (1–3):15–24. doi: 10.1016/s0168-3659(01)00486-2.

Vinikoor, L. C., J. C. Schroeder, R. C. Millikan, J. A. Satia, C. F. Martin, J. Ibrahim, J. A. Galanko, and R. S. Sandler. 2008. Consumption of trans-fatty acid and its association with colorectal adenomas. *American Journal of Epidemiology* 168 (3): 289–297. doi: 10.1093/Aje/Kwn134.

Vodyanoy, V. 2010. Zinc nanoparticles interact with olfactory receptor neurons. *Biometals* 23 (6):1097–1103. doi: 10.1007/s10534-010-9355-8.

Walle, T. 2011. Bioavailability of resveratrol. *Resveratrol and Health* 1215:9–15. doi: 10.1111/j.1749-6632.2010.05842.x.

Webster, J. L., E. K. Dunford, C. Hawkes, and B. C. Neal. 2011. Salt reduction initiatives around the world. *Journal of Hypertension* 29 (6):1043–1050. doi: 10.1097/Hjh.0b013e328345ed83.

Weltzin, R., P. Luciajandris, P. Michetti, B. N. Fields, J. P. Kraehenbuhl, and M. R. Neutra. 1989. Binding and trans-epithelial transport of immunoglobulins by intestinal M-cells—Demonstration using monoclonal IgA antibodies against enteric viral-proteins. *Journal of Cell Biology* 108 (5):1673–1685.

Wigginton, N. S., A. De Titta, F. Piccapietra, J. Dobias, V. J. Nesatty, M. J. F. Suter, and R. Bernier-Latmani. 2010. Binding of silver nanoparticles to bacterial proteins depends on surface modifications and inhibits enzymatic activity. *Environmental Science & Technology* 44 (6):2163–2168. doi: 10.1021/es903187s.

Xu, Q. A., X. Y. Yin, M. Wang, H. F. Wang, N. P. Zhang, Y. Y. Shen, S. Xu, L. Zhang, and Z. Z. Gu. 2010. Analysis of phthalate migration from plastic containers to packaged cooking oil and mineral water. *Journal of Agricultural and Food Chemistry* 58 (21):11311–11317. doi: 10.1021/jf102821h.

Yamada, Y. 1994. *Modern Textbook of Histology, Third Edition.* Tokyo: Kanehara & Co., Ltd.

Yamago, S., H. Tokuyama, E. Nakamura, K. Kikuchi, S. Kananishi, K. Sueki, H. Nakahara, S. Enomoto, and F. Ambe. 1995. In vivo biological behavior of a water-miscible fullerene-C-14 labelling, absorption, distribution, excretion and acute toxicity. *Chemistry & Biology* 2 (6):385–389. doi: 10.1016/1074-5521(95)90219-8.

Yang, H., L. W. Qu, A. N. Wimbrow, X. P. Jiang, and Y. P. Sun. 2007. Rapid detection of *Listeria monocytogenes* by nanoparticle-based immunomagnetic separation and real-time PCR. *International Journal of Food Microbiology* 118 (2):132–138. doi: 10.1016/j.ijfoodmicro.2007.06.019.

Zhang, Y. A., X. Q. Wang, J. C. Wang, X. A. Zhang, and Q. A. Zhang. 2011. Octreotide-modified polymeric micelles as potential carriers for targeted docetaxel delivery to somatostatin receptor overexpressing tumor cells. *Pharmaceutical Research* 28 (5):1167–1178. doi: 10.1007/s11095-011-0381-1.

Zimet, P. and Y. D. Livney. 2009. β-Lactoglobulin and its nanocomplexes with pectin as vehicles for ω-3 polyunsaturated fatty acids. *Food Hydrocolloids* 23 (4):1120–1126. doi: 10.1016/j.foodhyd.2008.10.008.

Zimet, P., D. Rosenberg, and Y. D. Livney. 2011. Re-assembled casein micelles and casein nanoparticles as nano-vehicles for ω-3 polyunsaturated fatty acids. *Food Hydrocolloids* 25 (5):1270–1276. doi: 10.1016/j.foodhyd.2010.11.025.

5

The Integumentary System
and the Ocular Route

Nanoproducts for administration via the integumentary system include nanomedicines, nanocosmetics/cosmeceuticals, nanoconsumer goods, and numerous products intended for delivery via the skin and hair. Nanoproducts for administration via the ocular route include nanomedicines and nanoconsumer goods. Dermal and ocular tissues are both affected by sunlight, as are nanoscale and conventional drugs and cosmetics/cosmeceuticals administered via the integumentary system and the eye. The disposition of nanoproducts and conventional products also changes due to photostability.

This chapter addresses the anatomy and physiology of the integumentary system and the eye, including the delivery, disposition, and practical application of relevant nanoproducts; nanoproduct photobiology, photostability, and phototoxicity; the intravenous route for the ocular delivery of nanoproducts; and the nanoscale world in terms of the integumentary system and the eye.

5.1 Anatomy and Physiology

5.1.1 Integumentary System

The integumentary system comprises the skin and its accessory organs. The skin is stratified into the epidermis, dermis, and hypodermis (also known as the subcutaneous layer). Accessory organs include the hair follicles and hair; the apocrine, eccrine, and sebaceous glands; the mammary glands; and the nails.

5.1.2 Dermal Cells

Dermal cells consist of keratinocytes, melanocytes, Langerhans cells, and Merkel cells (Figure 5.1). Keratinocytes undergo differentiation into surface (epidermal) or basal skin and are stratified into four layers containing (1) dead keratinized cells on the skin surface, (2) lamellar body-releasing granular cells, (3) spinous cells, and (4) basal cells. Melanocytes release melanin to the keratinized layer, while antigen-presenting Langerhans cells are

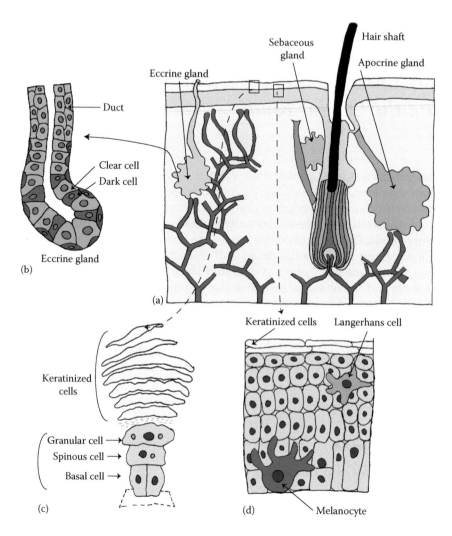

FIGURE 5.1

(a) Integumentary system, consisting of the skin and accessory glands, (b) eccrine gland consisting of dark and clear cells, (c) keratinocytes stratified into four layers, (d) keratinocytes including melanocytes and merkel cells. (a and d: Modified from Ross, M.H. and Pawlina, W., *Histology: A Text and Atlas*, 6th edn., Lippincott Williams & Wilkins, Baltimore, MD, 2011; b: Modified from Fujita, H. and Fujita, T., *Textbook of Histology: Part 2*, 3rd edn., Igaku-Shoin, Ltd., Tokyo, Japan, 1992; c: Modified from Squier, C.A. and Kremer, M.J., *J. Natl. Cancer Inst. Monogr.*, 29, 7, 2001.)

required for cutaneous adaptive immune responses. Merkel cells function as neurosensory cells that provide touch information to the brain.

Keratinocytes are ~8–16 μm in diameter and are stacked 17–30 layers high, depending on the area of body and individual deviation (Holbrook and Odland, 1974). Water volume increases with tissue thickness, from the outer keratinized layer to the basal layer (Warner et al., 1988; Caspers et al., 2000; Boncheva et al., 2009). Water accounts for ~15% of the volume in the first layer of keratinized cells in the freeze-dried dermis, progressively increasing to ~40%, 40%–70%, and 70% in the final keratinized layer, granular layer, and basal layer, respectively (Warner et al., 1988). The pH of the skin is also location dependent and changes from weakly acidic (pH 5) in the keratinized and granular layers to neutral in the basal layer (Matousek and Campbell, 2002). Lastly, most capillaries are found at a depth of 100 µm from the skin surface (Braverman, 1997), with relatively few at 750 µm (Cevc and Vierl, 2007).

Keratinized cells show an elasticity coefficient of 107–108 Pa, which increases with increasing temperature and hydration (Papir et al., 1975). The thickness of the tight junction space in the keratinized surface layer is 100 nm or less in normal mammalian skin (Fartasch et al., 1993), and the nanostructure of the epidermal extracellular space is revealed by full hydration (Al-Amoudi et al., 2005). The extracellular space of the viable epidermis contains desmosomes with a characteristic extracellular transverse periodicity of ~5 nm, whereas the extracellular space between the viable and cornified epidermis contains transition desmosomes. In addition, the extracellular space of the cornified epidermis contains several regions with a thickness of ~9, 4, 25, 33, 44, and 48 nm, which in turn contain 1, 2, 4, 6, 8, or 10 parallel electron-dense lines. The 44 nm thick, eight-line region is the most prevalent.

The ratio of Langerhans cells to keratinocytes in human breast skin is 1:53, while the corresponding ratio in mouse ear skin is 1:15 (Mulholland et al., 2006). The number of Langerhans cells in mice is three times higher than in humans.

5.1.3 Hair and Hair Follicles

Hair follicles are surrounded by arrector pili muscles and sebaceous glands. The hair follicle together with its associated hair is termed a follicular or pilosebaceous unit (Figure 5.2a). There are four types of pilosebaceous units in humans, including the terminal pilosebaceous, vellus pilosebaceous, apopilosebaceous, and sebaceous pilosebaceous unit (Kealey et al., 1997). The terminal pilosebaceous unit is located in the scalp of both genders and in the beard region of men. The vellus pilosebaceous unit is located in areas that appear hairless to the naked eye but are in fact covered with fine hairs. The apopilosebaceous unit is located in the axilla in association with apocrine glands, and the sebaceous pilosebaceous unit is located on the face, chest, and back.

Hair follicles can be either dermal or epidermal, depending on the distance from the skin surface. Dermal hair follicles comprise internal and external root sheaths. The internal root sheath is composed of three layers, Huxley's layer, Henley's layer, and an internal cuticle that is continuous with

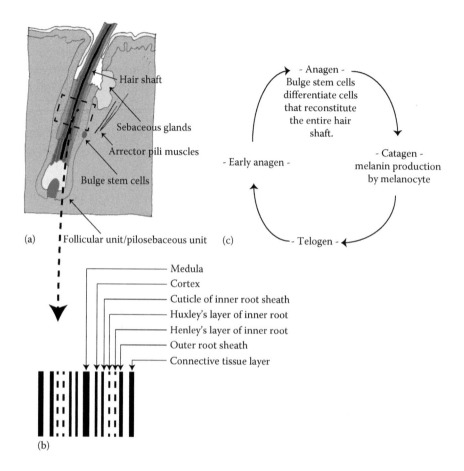

FIGURE 5.2
Pilosebaceous units. (a) Representative pilosebaceous unit. (b) Structure of the hair shaft.
(c) Life cycle of human hair.

the outermost layer of the hair fiber (Figure 5.2b). The external root sheath
contains stem cells for hair growth. Epidermal hair follicles comprise inner
and outer circular layers (Birbeck and Mercer, 1957a,b,c; Araujo et al., 2011).

Hair fibers consist of the hair shaft, or the cuticle, the cortex, and the
medulla. The growth rate of scalp hair is ~0.3 mm/day. Adult human hair
undergoes a periodic cycle of active growth, follicle shrinkage, and rest, corresponding to the anagen, catagen, and telogen phases, respectively (Figure
5.2c). The sebaceous pilosebaceous unit has a long periodicity of 4 years,
with 90% of the cycle spent in the anagen phase and 10% in the catagen and
telogen phases, while the vellus pilosebaceous unit has a short periodicity,
with an anagen phase of 54 days in males and 27 or less days in females in
the femoral area (Kealey et al., 1997).

5.1.4 Sweat Glands

Sweat glands are coiled, tubular structures that include the eccrine and apocrine sweat glands for temperature control and pheromone release, respectively. The eccrine sweat glands are composed of clear and dark secretion cells, myoepithelial cells, and a stratified cuboidal epithelium (Figure 5.1b). The apocrine sweat glands are composed of single-secretion epithelial cells, myoepithelial cells, and a stratified cuboidal epithelium.

5.1.5 Nails

Nails are composed of keratinized cells that produce a hardened keratin matrix. The nail root is located in the proximal area of nail and embedded in the dermis, surrounded by cells in the germinative zone. These cells include stem cells, epithelial cells, melanocytes, Merkel cells, and Langerhans cells. Stem cells migrate to the nail root, differentiate into keratinized cells, and release the keratin matrix. The hardness of the nail depends on close associations between keratin filaments in the matrix and high matrix concentrations of sulfhydryl-containing compounds.

5.1.6 Anatomy of the Eye

The eye comprises the eyeball, the eye chamber, and the accessory ocular organs. The eyeball is composed of three layers: the corneoscleral coat (Figure 5.3, blue-colored area, except for colorless area in cornea), the vascular coat/uvea (Figure 5.3, red), and the retina (also known as the tunica interna bulbi; Figure 5.3, yellow). The corneoscleral coat consists of the white sclera and the clear cornea; the vascular coat/uvea consists of the choroid, ciliary body, and iris; and the retina, which is linked by the optic nerve to the brain, consists of the outer neural layer, the photoreceptor layer, the retinal pigment epithelium, and melanin.

Like the eyeball, the eye chamber is composed of three parts: the anterior chamber (the space between the cornea and the iris), the posterior chamber (the space between the posterior surface of the iris and the anterior surface of the lens), and the vitreous chamber (the space between the posterior surface of the lens and the retina) (Figure 5.3). The accessory ocular organs include the conjunctiva, eyelid (not shown in Figure 5.3), and lacrimal organ (not shown in Figure 5.3).

5.1.7 Corneal and Scleral Cells

The corneoscleral coat contains the cornea (the transparent front part of the eye) and the sclera (the white of the eye). The cornea is composed of three cellular layers and two extracellular layers and includes (in order from the upper surface) the corneal epithelium, Bowman's membrane (also called the anterior basement membrane), the corneal stroma, Descemet's membrane

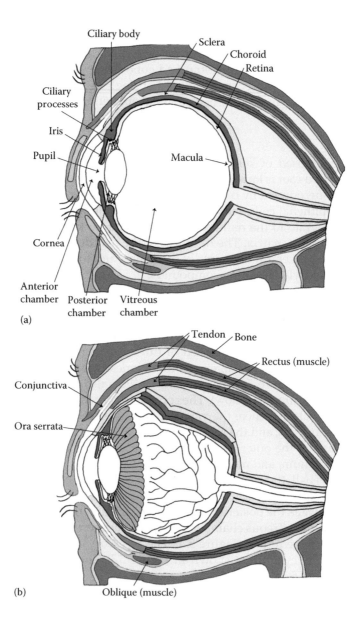

FIGURE 5.3
Structure of the eye. (a) Three layers of the eye. (b) Internal structure of the eye, showing the outer supporting layer and corneoscleral coat (colorless and blue, respectively), the middle vascular coat or uvea (orange), and the retina (yellow). (Modified from Ross, M.H. and Pawlina, W., *Histology: A Text and Atlas*, 6th edn., Lippincott Williams & Wilkins, Baltimore, MD, 2011; Standring, S., *Gray's Anatomy*, 40th edn., Churchill Livingstone, Elsevier, London, 2008.)

(the posterior basement membrane), and the corneal endothelium. The cornea has an average thickness of about 0.5 and 1 mm in the center and peripheral area, respectively.

The corneal epithelium has an average thickness of 50 μm and a turnover time of 7 days and contains five layers of nonkeratinized cells. The corneal epithelium grows in basal areas and expresses the iron storage protein ferritin, which protects DNA from ultraviolet (UV) damage. The 8–10 μm thick Bowman's membrane underlies the corneal epithelium and is made of laminar collagen fibrils (mean diameter, 18 nm) presented in a discrete orientation.

The corneal stroma accounts for 90% of the thickness of the cornea and comprises ~60 lamellae (flattened plates of collagen fibrils) that are 23 nm wide and 1 cm long. The lamellae are maintained in association with corneal proteoglycans, mainly lumican, a keratan sulfate proteoglycan, and decorin, a chondroitin sulfate proteoglycan. Lumican is essential for the regulation of ordered stromal collagen matrix assembly and corneal transparency and provides uniform spacing between the orthogonal arrays of collagen fibers. Lastly, Descemet's membrane (~10 μm thick) lies between the corneal stroma and endothelium and forms the basement membrane of the corneal endothelial cells.

The sclera is perforated by blood vessels and nerves and contains three opaque layers: the episcleral layer or episclera, the substantia propria (the sclera proper or Tenon's capsule), and the suprachoroid lamina or lamina fusca. The episcleral layer consists of loose connective tissue. The substantia propria forms the fascia of the eye and consists of a thick, dense collagen fibril network. The suprachoroid lamina overlies the choroid and consists of thin collagen fibrils, elastin fibrils, fibroblasts, melanocytes, macrophages, and other connective tissue cells.

The clear cornea meets the white sclera in the limbal transitional zone (Figure 5.4). There, ciliary processes secrete an aqueous fluid where the eyeball borders the lens in the posterior chamber. The aqueous fluid flows to the anterior chamber and is absorbed by the scleral venous sinus (the canal of Schlemm) via the spaces of Fontana. The pores in the canal of Schlemm are 0.5–1.5 μm in diameter in humans (Johnson et al., 2002), and therefore 500 nm particles pass through the outflow pathway of enucleated human eyes at a frequency of 50% (Johnson et al., 1990).

5.1.8 Cells in the Uvea or Vascular Coat

The iris within the uvea contains concentric folds and exerts anterior force to change the size of the central pupil in response to light stimulation. The size of the pupil is controlled by sphincter papillae and dilator papillae muscles. The iris also contains an abundant supply of blood vessels, a loose fibrocollagenous stroma, and a posterior pigment epithelial coating.

The ciliary body extends ~6 mm from the basal part of the iris to the ora serrata, which contains a 17–34 μm groove. One-third of the anterior ciliary

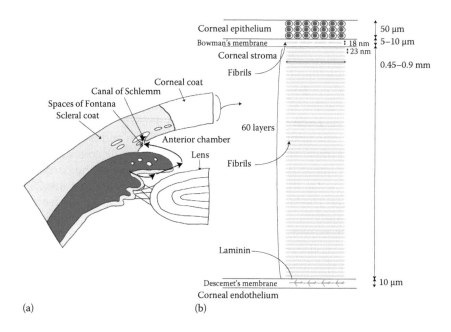

FIGURE 5.4

Structure of the eyeball, corneal coat, scleral coat, and conjunctiva. (a) Sagittal section of the eyeball, cornea coat, scleral coat, and conjunctiva. (b) Structure of the cornea coat. (Modified from Yamada, Y., *Modern Textbook of Histology*, 3rd edn., Kanehara & Co., Ltd., Tokyo, Japan, 1994; Ross, M.H. and Pawlina, W., *Histology: A Text and Atlas*, 5th edn., Lippincott Williams & Wilkins, Baltimore, MD, 2006; Standring, S., *Gray's Anatomy*, 40th edn., Churchill Livingstone, Elsevier, London, 2008.)

body is covered with 75 radially extending ciliary processes. The ciliary body exhibits a layered structure that is similar to that of the iris, with a stroma and an epithelial coating. The ciliary stroma includes the outer ciliary smooth muscle and the inner ciliary processes. The ciliary epithelium secretes aqueous humor, maintains the blood–ocular barrier, and secretes and anchors suspensory ligament-containing zonular fibers.

The choroid within the uvea is located in the deep part of the retina, between the retina and the sclera. The choroid is a dark brown, thin vascular plate with a posterior thickness of 0.25 mm and an anterior thickness of 0.1 mm. TEM investigation of the choroid using freeze-fracture replication showed that the size of the fenestrae in the choriocapillaries ranged from 75 to 85 nm in human donor eyes and monkey and hamster eyes (Guymer et al., 2004).

5.1.9 Reticular Cells

The retina consists of the retinal pigment epithelium and the multilayered neural retina (Figure 5.5). The neural retina includes photosensitive

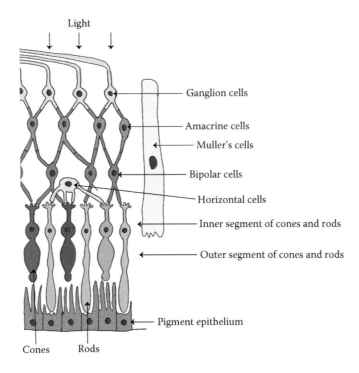

FIGURE 5.5
Structure of the retina. (Modified from Ross, M.H. and Pawlina, W., *Histology: A Text and Atlas*, 5th edn., Lippincott Williams & Wilkins, Baltimore, MD, 2006; Standring, S., *Gray's Anatomy*, 40th edn., Churchill Livingstone, Elsevier, London, 2008.)

regions (containing rods and three types of cone cells) and nonphotosensitive regions. The optic disk or optic papilla is located where the optic nerve joins the neural retina.

The retinal pigment epithelium is composed of a single, 14 µm thick, and 10–14 µm wide layer of columnar cells. The blood–retinal barrier is formed by direct cell-to-cell contact between retinal pigment epithelial cells and nonfenestrated capillaries via junctional complexes. The outer nuclear layer of the neural retina lies underneath the retinal epithelium and contains a layer of rods and cones. Rods are highly sensitive to light and respond to gray-tone images (night vision), while cones have low sensitivity to light and absorb red, green, or blue stimuli (day vision), depending on the type of cone present.

The outer limiting membrane forms a metabolic barrier to the inner retina by limiting the passage of high-molecular-weight (HMW) substances. The outer limiting membrane is situated dorsal to the outer nuclear layer and ventral to the outer plexiform layer of neuronal synapses. The inner nuclear layer follows and is composed of bipolar cells, horizontal cells, amacrine cells, interplexiform cells, and Müller cells, forming the base of the entire retina. Next, the inner plexiform layer contains the axons of bipolar neurons and the dendrites of retinal ganglion cells, where the ganglion cell layer contains the 30 μm cell somata of large, multipolar ganglion neurons and the optic nerve layer contains the ganglion cells axons.

Finally, the inner limiting membrane forms a basement membrane boundary between the retina and the vitreous body, and the fovea centralis forms a 200 μm depression in the center of the retina that is made only of cones and constitutes the area of most acute vision.

5.2 Practical Applications in Skin Care

Nanoproducts utilized for practical application in the integumentary system include sunscreens for safeguarding the skin against UV radiation, products for appearance enhancement, and assorted antimicrobial, skin care, hair care, dental care, and nail care products. This chapter focuses on sunscreens incorporating nanomaterials and nanoproducts for appearance enhancement of the skin.

5.2.1 Sunscreens

Excessive exposure of the skin to UV radiation in the form of sunlight induces adverse events, including sunburn, premature aging, immunosuppression, and cancer. However, complete avoidance of the sun is impractical, and reasonable sun exposure also contributes to protection against disease. Therefore, sunscreen agents afford one of the best strategies for protecting the skin from disproportionate UV radiation.

In Europe, the four predominant indicators for evaluating the efficacy of a sunscreen agent comprise the sun protection factor, the UVA protection factor (PF-UVA), the SPF/PF-UVA ratio, and the critical wave length (Couteau et al., 2012). A truly efficacious product protects against sunburn and nonmelanocyte skin cancer. Sehedic et al. (2009) evaluated 35 products in the market according to SPF/PF-UVA ratio and critical wavelength. Seven of these products (20%) failed to adequately protect against UV radiation, and 27 products (77%) demonstrated instability against sunlight. All of the marketed products met the recent standards of an SPF/PF-UVA ratio of ≤3 and a critical wavelength of ≥370 nm, and all showed a protective effect against UV radiation, albeit mainly against UVB rays. However, they did not sufficiently safeguard the skin from UVA rays, which penetrate more deeply into the skin than UVB rays.

Nanoproducts for use as sunscreens include metallic and lipid nanoparticles, as well as carbon and dendrimer nanomaterials. Metallic sunscreens contain titanium dioxide and zinc oxide nanoparticles characterized by (1) 5–20 nm primary particles on the date of manufacture, (2) 30–150 nm nanoparticle aggregates generated by attractive chemical forces, and

(3) 1–100 µm nanoparticle agglomerates generated by van der Waals forces (Figure 5.6). In addition to these aforementioned characteristics of inorganic sunscreens, organic sunscreens retain the space of applications in the future, although the sunscreen products are limited to inorganic particles in the current market. Organic sunscreens are characterized by (1) nanoparticles generated by physical interaction and (2) nanoparticles generated by electrostatic forces (Figure 5.6). The preponderance of inorganic primary particles versus nanoparticle aggregates or agglomerates affects the protective impact of the sunscreen against UV rays (Wang and Tooley, 2011). The importance of optimal particle size is exemplified in Figure 5.7. Titanium dioxide nanoparticles with a diameter of 100 nm absorb UVA and UVB rays and also visible light. However, 50 nm titanium dioxide nanoparticles show a high absorptive capacity against UVB rays but not against UVA rays, and 20 nm nanoparticles show low absorptive capacity against both UVA and UVB rays.

5.2.2 Appearance Enhancement

"Nutricosmetics" is a new concept used to describe foods or oral supplements employed for appearance enhancement. These consumer goods are mainly marketed as beauty or beauty-from-within pills, oral cosmetics (Draelos, 2010), and antiaging products for the reduction of wrinkles caused by sun-induced free radical formation. Examples include carotenoids (β-carotene, lycopene, lutein, zeaxanthin, and astaxanthin) and polyphenols (anthocyanidins, catechins, epigallocatechin-3-gallate, flavonoids, tannins, and procyanidins). All of these compounds exhibit free radical scavenging properties (Anunciato and da Rocha, 2012). Whitening nutricosmetics have also been developed that target melanin production by melanocytes for the lightening of melasmas.

Tyrosine kinase is activated during UV-induced melanogenesis and, therefore, inhibition of tyrosine kinase is the predominant mechanism for the downregulation of melanin production. Other mechanisms include inhibition of melanosomal transfer, acceleration of epidermal turnover and desquamation, promotion of antioxidant activity, and delivery of conjugates to improve the stability and efficacy of skin-lightening agents (Gillbro and Olsson, 2011). Melanin is synthesized in the melanocyte melanosome as either eumelanin, a dark brown insoluble polymer, or pheomelanin, a slightly red-yellow soluble polymer (Ito and Wakamatsu, 2003). Only eumelanin has the ability to remove free radical molecules and to bind amino acids and heavy metals (Larsson, 1993; Dunford et al., 1995; Rozanowska et al., 1999).

Metallic nanoparticles, nonmetallic nanoparticles, nanoclays, and synthetic mesoporous silica nanomaterials have all been investigated for their capacity to act as nutricosmetics. A 5–300 nm retinoic acid–loaded nanoparticle has recently shown good efficacy against skin wrinkling when delivered via the dermal route (Yamaguchi et al., 2006). Retinoic acid stimulates cell differentiation and decreases skin wrinkling by heightening the turnover time of dermal keratinocytes.

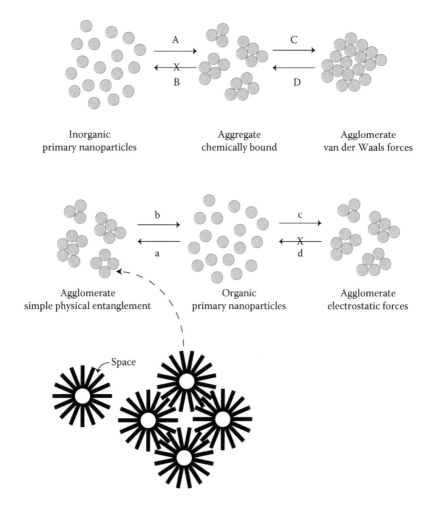

FIGURE 5.6
Primary nanoparticles, aggregated nanoparticles, and agglomerated nanoparticles. (A) Because of the strong attractive forces between inorganic primary nanoparticles, the particles cluster together and form tightly bound aggregates, which have a larger size than their primary building blocks. (B) The attractive forces discourage dissociation of the aggregates. (C) The aggregates form loosely bound agglomerates because of the drying and heat-treatment processes employed during manufacturing. (D) Large nanoparticle agglomerates do not provide sufficient protection against UV radiation. Therefore, the agglomerates must be broken down into smaller aggregates. (a) Because of the simple physical entanglement between organic primary nanoparticles during manufacturing, the particles cluster together and form weakly bound agglomerates, which have a larger size than their primary building blocks. (b) The agglomerate return primary nanoparticle simple physical entanglements discourage dissociation of the agglomerates, because of diffusion. (c) Because of the electrostatic forces between primary nanoparticles, the particles cluster together and form weakly bound agglomerates, which have a larger size than their primary building blocks. (d) The weak forces discourage dissociation of the agglomerates.

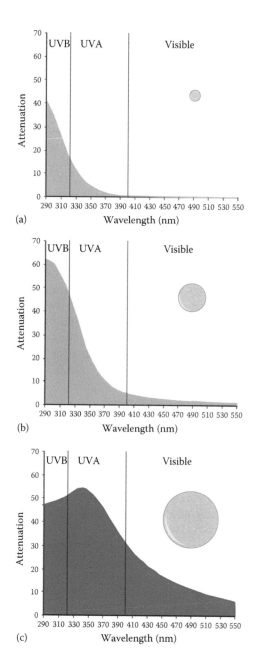

FIGURE 5.7
Attenuation (absorption) of UV and visible light by spherical titanium dioxide nanoparticles as a function of radiation wavelength and particle size. (a) 20 nm primary particles; (b) 50 nm aggregated particles; and (c) 100 nm agglomerated particles. (Modified from Wang, S.Q. and Tooley, I.R., *Semin. Cutan. Med. Surg.*, 30(4), 210, 2011, doi: 10.1016/j.sder.2011.07.006.)

5.3 Practical Applications in Ocular Medical Care

The improvement of bioavailability in drug delivery is a primary concern in ocular medical care (Ali and Lehmussaari, 2006; Sahoo et al., 2008; Gaudana et al., 2009; Diebold and Calonge, 2010; Liu et al., 2012). Anatomically, the ocular route includes the anterior, posterior, and vitreous chambers, the aqueous and vitreous humors, and the pupil, cornea, iris, ciliary body, lens, retina, choroid, macula, and optic nerve (Figure 5.8). Therapeutically, the ocular route includes the intravenous, topical, periocular, and intravitreal/intraocular pathways (Figure 5.8). Vitreous humor has a volume of 3.9–5.0 mL in humans, while both the aqueous and vitreous humors undergo one-way circulation at a rate of 2.5–3 μL/min, followed by elimination (Meredith, 2006).

The predominant eye diseases encompass age-related macular degeneration (AMD), diabetic macular edema, cataracts, proliferative vitreoretinopathy, uveitis, cytomegalovirus, and glaucoma. A simulation study of systemic flow after injection of memantine, a centrally acting drug, into fresh bovine eyes demonstrated that drug distribution differed for each route of administration (Koeberle et al., 2006). Nanotechnological advances have now permitted the development of new therapeutic agents for delivery via the various ocular routes (Gaudana et al., 2009; Liu et al., 2012).

5.3.1 Nanoproduct Delivery via the Intravenous Route

Issues for the ocular delivery of conventional drugs include overcoming the lack of pores in iris endothelial cells, limiting the penetration of drugs into the interior of the eye via the blood–aqueous barrier of the nonpigmented layer of the ciliary epithelium, and limiting the penetration of drugs from the blood into the retina via the blood–retinal barrier. An intraocular circulation model using memantine showed that conventional drugs preferentially circulated to the ciliary body via blood vessels after intravenous injection. In this model, drugs first distributed to the anterior aqueous humor in high concentrations, followed by the cornea and retina, and, finally, to the posterior vitreous humor in low concentrations (Figure 5.9) (Koeberle et al., 2006). Because only 1%–2% of the injected volume distributed to the vitreous cavity, the intravenous route is clearly not optimal for delivery to the vitreous chamber. On the other hand, nanoproducts easily distribute to sites of inflammation because of enhanced angiogenesis, as discussed in Chapter 2. Therefore, the intravenous route enables nanomedicine therapy of ocular inflammation (Mousa et al., 2007; Mohan et al., 2012).

AMD induces the development of abnormal neovasculature in the choroid and also in the retina as the disease progresses. However, blood vessels are weak in AMD, and AMD patients experience remarkably decreased vision secondary to hemorrhage, exudate storage, abnormal function of the

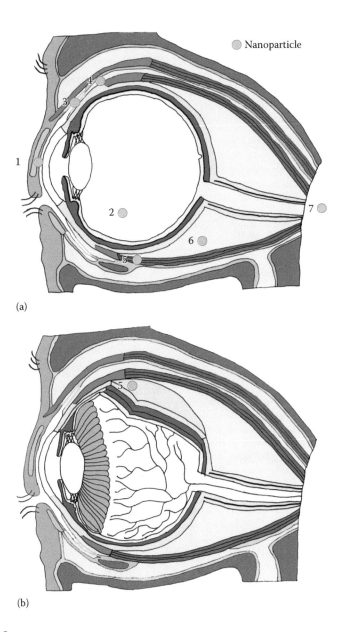

FIGURE 5.8
Route of administration in ocular delivery system for conventional products and nanoproducts: (a) route of administration in three layers of the eye and (b) route of administration in internal structure of the eye. (1) Topical eyedrop; (2) intravitreal injection; (3) periocular injection, subconjunctiva; (4) periocular injection, peribulbar; (5) periocular injection, subtenon; (6) periocular injection, retrobulbar; and (7) intravenous injection. (Modified from Standring, S., *Gray's Anatomy*, 40th edn., Churchill Livingstone, Elsevier, London, 2008; Gaudana, R. et al., *Pharmaceut. Res.*, 26(5), 1197, 2009, doi: 10.1007/s11095-008-9694-0; Liu, S.Y. et al., *Macromol. Biosci.*, 12(5), 608, 2012, doi: 10.1002/mabi.201100419.)

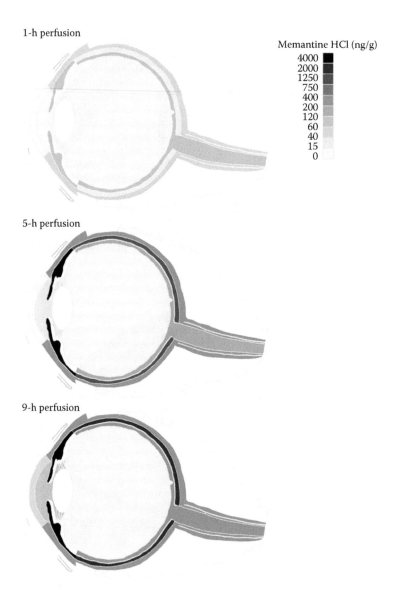

FIGURE 5.9
Drug distribution via the intravenous route. Dosage and dose of intravenous route (1.0 μM, 1.0 mL/min, 12.9 μg/h). The lens, conjunctiva, and optic nerve were not analyzed by Koeberle et al. (Modified from Koeberle, M.J. et al., *Pharmaceut. Res.*, 23(12), 2781, 2006, doi: 10.1007/s11095-006-9106-2.)

macular area, and, ultimately, irreversible macular degeneration. The main therapeutic methods for removing abnormal new blood vessels are photodynamic therapy with Visudyne® (Bressler and Bressler, 2000; Frennesson and Nilsson, 2004) and chemotherapy with antiangiogenetic agents.

Visudyne is a liposome formulation loaded with the photosensitive drug verteporfin. Verteporfin is activated by light in the presence of oxygen and releases highly reactive, short-lived singlet oxygen and reactive oxygen radicals. Five minutes after a 10 min intravenous infusion of Visudyne, neovasculature is destroyed by irradiation with an extremely weak laser emitting 689 nm wavelength light. Unfortunately, this therapeutic strategy only delays the onset of blindness and cannot cure AMD.

5.3.2 Topical Eyedrops

The use of topical eyedrops employs two drug delivery routes: one via the cornea and the other via the conjunctiva. Conventional drugs administered in this manner can be lost to the tears by lacrimal drainage or drug dilution, and drug penetration is limited by the anatomical properties of the epithelium, propria, and endothelium. For example, eyedrops added to the surface of the eye are washed out within 5–6 min (Kaur and Kanwar, 2002). Furthermore, the intraocular memantine circulation model indicated that drugs administered via topical eyedrops distribute to the anterior chamber (sclera, iris, and ciliary body) at high levels, circulate to the posterior chamber, and then distribute to the choroid near the vitreous body and into the vitreous humor at low levels (Figure 5.10) (Koeberle et al., 2006).

Tight junctions and desmosomes in the corneal epithelium restrict the transport of substances between cells. Even conventional drugs of ≥500 Da do not pass through the corneal epithelium via paracellular transport in rabbits (Hamalainen et al., 1997). The cutoff size for passage through the paracellular space in the corneal epithelium was 2 nm, whereas the corresponding cutoff size in the palpebral and bulbar conjunctiva was 4.9 and 3.0 nm, respectively (Nagel et al., 2003). A transport study of the topical eyedrop administration of timolol maleate showed that the low-molecular-weight (LMW) drug distributed to the conjunctiva at high levels and then traveled to the cornea, sclera, iris–ciliary body, aqueous humor, lens, and vitreous humor (Ahamed and Patton, 1985). These results suggest that the conjunctiva provides an important route for drug administration in addition to the corneal route. Finally, distribution after topical eyedrop delivery of conventional drugs is only 5% of the injected volume in the aqueous humor (Ahamed and Patton, 1985; Keister et al., 1991).

5.3.3 Nanomedicine Releasability and Rigidity

Drug delivery of nanoproducts versus conventional drugs for topical eyedrop administration can potentially increase the adhesion time of drugs

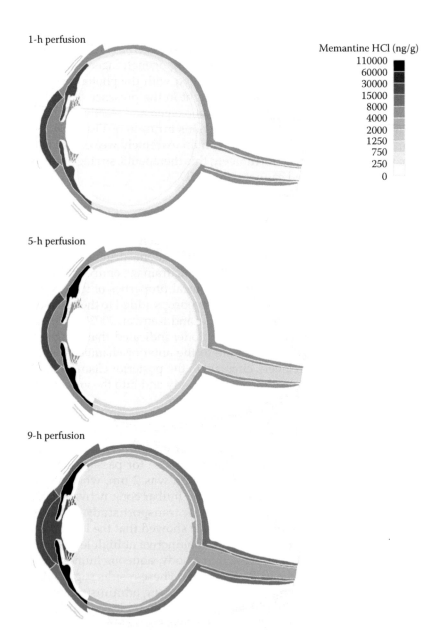

FIGURE 5.10
Drug distribution via topical eyedrops. Dosage and dose of topical eyedrops (9.27 μM, 4.0 mL topical reservoir, 8.002 μg/reservoir). The lens, conjunctiva, and optic nerve were not analyzed by Koeberle et al. (Modified from Koeberle, M.J. et al., *Pharmaceut. Res.*, 23(12), 2781, 2006, doi: 10.1007/s11095-006-9106-2.)

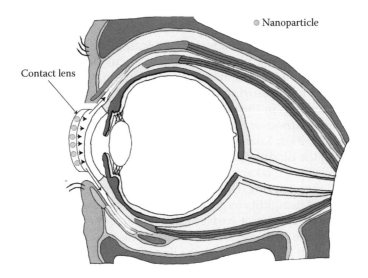

FIGURE 5.11
Ocular delivery via contact lens. (Modified from Gulsen, D. and Chauhan, A., *Int. J. Pharmaceut.*, 292(1–2), 95, 2005, doi: 10.1016/j.ijpharm.2004.11.033; Liu, S.Y. et al., *Macromol. Biosci.*, 12, 608, 2012, doi: 10.1002/mabi.201100419.)

with the surface of the eye (Gulsen and Chauhan, 2005; Inokuchi et al., 2010; Hironaka et al., 2011). Two strategies are employed for this purpose. The first uses biodegradable composites formulated as a contact lens containing drug-loaded nanoparticles to enable slow release of drug, penetration into the cornea or conjunctiva, circulation to the aqueous and vitreous humor, and distribution to the retina (Figure 5.11). The second strategy controls the rigidity of nanoparticles so that nanomedicines can penetrate into corneal and conjunctiva cells via endocytosis and distribute to the retina via the aqueous and vitreous humor (Figure 5.11).

5.3.4 Contact Lens Composites

The contact lens as a drug-loaded composite is a plate-shaped film that slowly releases the drug into tears found in the prelens (the film between the air and the lens) and the postlens (the film between the cornea and the lens) (Gulsen and Chauhan, 2005). To manufacture contact lens composites, a poly-2-hydroxyethyl methacrylate hydrogel is loaded with a microemulsion that is stabilized with a silica shell. The gel releases the drug for a period of 4 days, and thus, the contact lens is expected to deliver drugs at therapeutic levels for at least a few days. The size of the primary nanoparticle in the gel is ~15 nm in diameter, but the diameters vary from about 150 to 500 nm by aggregation during the water-evaporation step.

5.3.5 Nanoparticle Rigidity

The rigidity of nanoparticles affects the releasability of active ingredients, the stability of the drug platform, and the circulation time in the blood. However, until the advent of atomic force microscopy (AFM), there was no reliable method for measuring nanoparticle rigidity. AFM first enabled the detection of changes in the mechanical properties of biological substrates, such as surface stiffness or elasticity in cells and cellular membranes (Alonso and Goldmann, 2003; Kuznetsova et al., 2007). Recently, AFM was adapted for measuring the rigidity of nanomaterials, including polystyrene nanoparticles and liposomes (Nakano et al., 2008).

Dynamic light scattering (DLS) permits the size assessment of a nanoparticle, regardless of its rigidity as a soft or hard object. However, the absorption of a nanoparticle to a mica surface triggers force- and rigidity-dependent nanoparticle shape changes that are readily measured by AFM. Under these conditions, the circular shape of a hard liposome is almost unalterable, while that of a soft liposome morphs into a flat shape (Figure 5.12). From the height (H) of the particle measured by AFM and the particle size (P) measured by DLS, the H/P value signifies the rigidity of the nanoparticle. Namely, a high H/P value indicates that the nanoparticle is hard.

The influence of nanoparticle rigidity on uptake into the conjunctiva was explored by using two types of fluorescently labeled liposomes: an egg phosphatidylcholine (EPC) liposome and a L-α-distearoyl-phosphatidylcholine

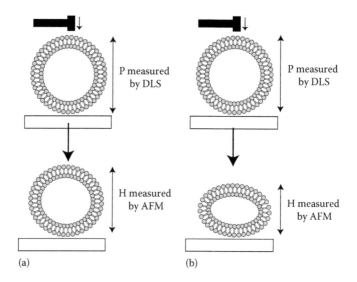

(a)   (b)

FIGURE 5.12
A method for measuring nanoparticle rigidity: (a) Hard- and (b) soft-type nanoparticles. (Modified from Nakano, K. et al., *Int. J. Pharm.*, 355, 203, doi: 10.1016/j.ijpharm.2007.12.018.)

(DSPC) liposome (Hironaka et al., 2009, 2011). The H/P values of the EPC and DSPC liposomes were 0.31 and 0.81, respectively, indicating that the DSPC liposome was harder than the EPC liposome. Uptake in conjunctiva cells was 50% or more for both liposomes (Hironaka et al., 2011), but uptake after topical eyedrop delivery of the DSPC liposome (105.4 nm) was higher than that of the EPC liposome (125.03 nm).

The impact of various physiochemical properties on the delivery of nanoparticles to the retina was explored after topical eyedrop administration in mice, rabbits, and monkeys (Inokuchi et al., 2010). The nanoparticles were loaded with fluorescently labeled coumarin-6, and the measured physiochemical properties included particle size, platform (liposome, lipid emulsion, polystyrene), rigidity by cholesterol content, surface charge, and monolayer versus multilayer structure. Surface charge (+25.9 mV = cationic, −3.0 mV = neutral, −53.9 mV = anionic) did not affect retinal delivery after 30 min. Neither 109 nm lipid emulsion particles nor 110 nm polystyrene particles traveled to the retina, but liposomes of 600 nm or less did, depending on the size decrease after 30 min. Soft liposomes with a low cholesterol content were also preferentially delivered to the retina.

Taken together, these findings suggest that soft liposomes have an advantage over hard liposomes for travel to the retina, while hard liposomes have an advantage for uptake into the cell.

5.3.6 Intraocular Injection

A study of intraocular/intravitreal injection using memantine showed the benefits of this route for retinal delivery (Koeberle et al., 2006). Memantine mainly distributed to the vitreous humor, traveled through the blood vessels to the choroid at high levels, traveled to the iris and ciliary body, and finally distributed to the anterior humor at low drug levels (Figure 5.13). Because an intraocular injection involves the vitreous cavity, this route enables the injection of high concentrations of drugs and the long-term retention of nanoparticles.

A study of intraocular injection underneath the retina in rabbits showed a size-dependent retention of nanoparticles. Fluorescent particles of 4 nm, which is smaller than the 7 nm size of albumin, were eliminated from the subretinal fluid after 30 h, while two fluorescent particles of 17 and 11.6 nm accumulated in the subretinal fluid after even 3 days (Marmor et al., 1985).

Macugen is a sterile, aqueous solution containing pegaptanib sodium for intravitreal injection that is indicated for the treatment of neovascular AMD. Pegaptanib sodium has a molecular weight of ~50 kDa and is likely a nanoscale compound rather than a nanoparticle. In clinical studies (Kourlas and Schiller, 2006), both Macugen (0.3 mg)-treated and sham-treated patients continued to experience vision loss. However, the rate of vision decline in the Macugen-treated group was slower than that in the sham-treated group.

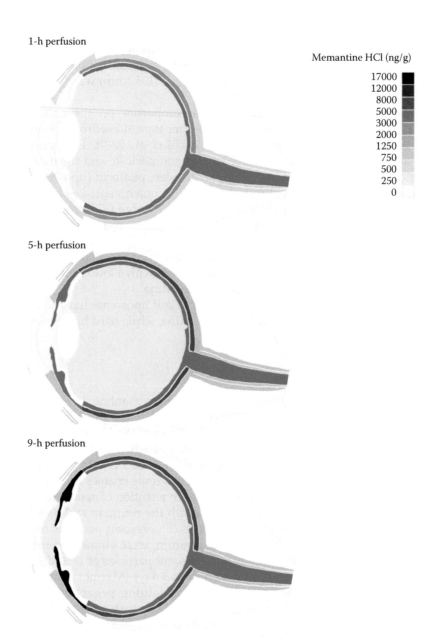

FIGURE 5.13
Drug distribution via intraocular injection. Dosage and dose of intraocular injection (250 μL, 397 μg). The lens, conjunctiva, and optic nerve were not analyzed by Koeberle et al. (Modified from Koeberle, M.J. et al., *Pharmaceut. Res.*, 23(12), 2781, 2006, doi: 10.1007/s11095-006-9106-2.)

5.3.7 Periocular Injection

Periocular injection enables drug distribution to both the anterior and posterior humors (Raghava et al., 2004; Ghate et al., 2007). Periocular injection makes use of the peribulbar, posterior juxtascleral, retrobulbar, subtenon, and subconjunctival routes near the sclera for drug administration. The sclera is mainly composed of fibrous tissue, and therefore drugs can easily pass through the tissue at a sustained high concentration near the retina and vitreous body (Raghava et al., 2004). Periocular injection of sodium fluoride via the subtenon in rabbits enabled higher retention in the posterior vitreous humor versus the anterior aqueous humor than periocular injection via the retrobulbar or subconjunctival pathway (Ghate et al., 2007).

Periocular injection shows variations in circulation to the retina via the vascular and lymphatic systems, and thus the design of nanoparticle size is critical (Amrite and Kompella, 2005). An intraocular injection study of 20 and 200 nm carboxyl-modified polystyrene nanoparticles indicated that the 20 nm nanoparticle traveled more extensively through the vasculature and the lymph tissue than the 200 nm particle, while the 200 nm particle accumulated in the retina for longer periods of time (Amrite and Kompella, 2005).

5.4 Exposure to Sunlight and Nanoparticle Toxicity

Administration of nanoproducts to the skin or eye results in exposure of the products to sunlight. Although free LMW compounds are often inactivated by the sun, the effects of sunlight may be ameliorated by the encapsulation of active ingredients in a nanoparticle. This is an important issue, given the warnings regarding the destruction of the earth's ozone layer and the resultant increase in UV radiation exposure from the sun.

The "Montreal Protocol on Substances that Deplete the Ozone Layer" was designed by the United Nations Environment Programme to reduce the production and consumption of ozone-destructive goods. The working group of the Montreal Protocol, the Environmental Effects Assessment Panel, reports the periodic effects of UV radiation due to sunlight on humans (Andrady et al., 2012). The frequency of skin cancer (melanoma and nonmelanoma forms) has continued to increase due to UVB exposure from the sun over the past 30 years. Furthermore, the rare type of cancer, Merkel cell carcinoma, is increasingly detected on sunlight-exposed portions of the skin. Moreover, in addition to conjunctiva melanoma located on the outside of the eye, melanoma located inside of the eye is now a risk of sun exposure.

The U.S. FDA Center for Drug Evaluation and Research published a study guideline for photosafety in 2003 for global finalization by the *International Conference on Harmonisation*. The aim of the photosafety guideline is to assess the potential of photoirritability and skin cancer linked with sunlight-mediated UVA and UVB radiations.

UVA and UVB radiations both augment the risk of cataract formation. UV radiation is a major risk factor for retinopathy, even in infants. Consistent wearing of sunglasses that screen out UV rays of 400 nm or less decreases the risk of early cataract formation and retinopathy. Short blue visible light (400–440 nm) is also a risk factor for retinopathy in adults (Roberts, 2011).

5.4.1 Photobiology of the Integumentary System

Sunlight comprises UV radiation (wavelength, 200–400 nm), visible radiation (400–760 nm), and infrared ray radiation (760 nm–10 μm) (Figure 5.14). UV radiation in turn consists of UVC (200–280 nm), UVB (280–315 nm), UVA (315–400 nm), UVAI (340–400 nm), and UVAII (315–340 nm) rays. UV, visible, and infrared ray radiations all exert effects on tissues and cells. Visible and infrared ray radiations mainly act to elevate temperature. UV radiation has contradictory actions in that UV light is beneficial for the production of vitamin D (Chen et al., 2007; Kimlin, 2008) and detrimental in terms of carcinogenicity (de Gruijl et al., 1993).

Different wavelengths of radiation penetrate human skin to various degrees. UVA rays can penetrate into deeper regions of the skin than UVB rays (Beissert and Granstein, 1996). Most (80%–100%) of the vitamin D in the body is synthesized from 7-dehydrocholesterol and requires exposure to sunlight in the 270–310 nm UVB region (Kimlin, 2008; Moan et al., 2012). Nonetheless, an animal study indicated that the risk of cancer was highest at a wavelength of ~293 nm, whereas the risk at 400 nm was almost nonexistent (de Gruijl et al., 1993). Another study showed that the number of p53 gene mutations linked with skin cancer was similar for UVA and UVB exposure (Halliday et al., 2005).

The frequency of cutaneous malignant melanoma (CMM) continues to rise in the Caucasian population. Meta-analyses of human CMM demonstrate that UVA radiation is the primary risk factor for sunlight-induced melanoma (Mitchell and Fernandez, 2012). An additional analysis was performed by the International Agency for Research on Cancer in 2009, and the ensuing reports demonstrated a link between the increased occurrence of CMM, augmented UVA exposure, and an effect on vitamin D metabolism via decreased UVB exposure.

Most (95%) of the UV energy reaching the surface of earth is UVA radiation, while UVB radiation accounts for ~5%. UVC light is the most harmful form of UV radiation but is scarcely relevant under normal conditions, and therefore UVA energy is responsible for most of the harmful effects of UV energy to humans (Timares et al., 2008; Pfeifer and Besaratinia, 2012; Moan et al., 2012). UVA light triggers the generation of reactive oxygen species and secondary lipoperoxidation, protein oxidation, and oxidative DNA damage. The latter materializes in the form of cyclobutane pyrimidine dimers and

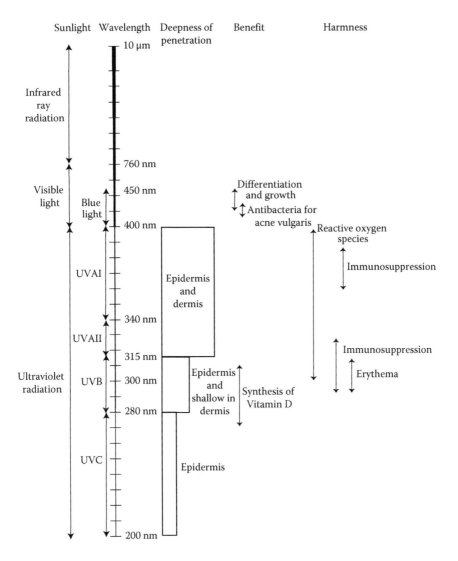

FIGURE 5.14
Sunlight and beneficial versus harmful effects on the skin.

8-oxo-7,8-dihydro-2-deoxyguanosine lesions (Timares et al., 2008; Pfeifer and Besaratinia, 2012).

Additional effects of UV radiation include wavelength-dependent immunosuppression (Halliday et al., 2011) and skin cancer. The induction of skin cancer following UV exposure is both wavelength and penetration depth dependent. The extent of penetration into the skin depends not only on the type of radiation but also on the amount of melanin released from melanocytes. For example, black people generate more melanin than white people, and melanocytes with high levels of melanin are more resistant against UVA-induced cytotoxicity than those with low levels of melanin (Yohn et al., 1992).

Sensitive wavelengths for immunosuppression show a biphasic pattern, with 364–385 nm as the range and 370 nm as the peak for UVA-induced immunosuppression and 290–320 nm as the range and 310 nm as the peak for UVB-induced immunosuppression (Matthews et al., 2010; Halliday et al., 2011).

Sunlight also induces effects at non-UV wavelengths in keratinocytes and skin-derived endothelial cells (Liebmann et al., 2010). For example, blue light at 412–426 nm affects skin cell differentiation and growth. Specifically, blue light triggers chromogenic effects via nitrosated heme proteins, initiating differentiation and inhibiting growth. Furthermore, *Propionibacterium acnes*, the bacterial source of acne vulgaris, absorbs blue light via porphyrin molecules, and the therapeutic effect of blue light against acne at ~415 nm is well known (Zanolli, 2003). Recently, Kleinpenning et al. (2010) studied the effects of a 5-day blue light exposure on photodamage to human skin associated with DNA modifications, normal skin aging, and melanogenesis. Although blue light induced melanogenesis, it did not cause photodamage or skin aging.

5.4.2 Photobiology of the Eye

Ocular sensitivity to sunlight damage in the cornea, lens, and retina depends on the radiation wavelength (Figure 5.15) (Dillon, 1991; Boulton et al., 2001; Hunter et al., 2012). The cornea is capable of absorbing wavelengths of 290 nm or less, and the most severe damage to the cornea occurs at a wavelength of 270 nm. The lens absorbs UVB rays at 300–315 nm and UVA rays at 315–400 nm. A photobiological study in rabbits showed that wavelengths of 300–305 nm caused the most pronounced damage to the lens, while an epidemiological study in humans indicated that wavelengths of 290–320 nm provoked corneal cataracts (Dillon, 1991). Hence, the two studies showed good agreement for ocular sensitivity to damaging wavelengths. In addition, a photobiological study in monkeys showed that wavelengths of 400 nm or less easily induced retinal damage.

The sclera and the lens absorb infrared radiation at 980, 1,200, and 1,430 nm, while the vitreous humor absorbs light at 1,400–100,000 nm. Therefore, the retina is exposed to radiation at 380–760 nm, corresponding mainly to visible light with electromagnetic properties, as well as to a portion of the infrared spectrum. The permeability of the cornea and lens to light changes with age in humans. The percentage of the area that is permeable to light at 320 nm is ~8% at 5 years of age and decreases to ~0.1% at 22 years of age. A larger percentage is permeable to light at 390 nm at a young age and reaches 90% at 450 nm, while the permeable region at 63 years of age increases at 400 nm but does not reach 90% even at 540 nm. Therefore, the retina of humans aged 60 years or more experiences enhanced exposure to blue light rather than UV light.

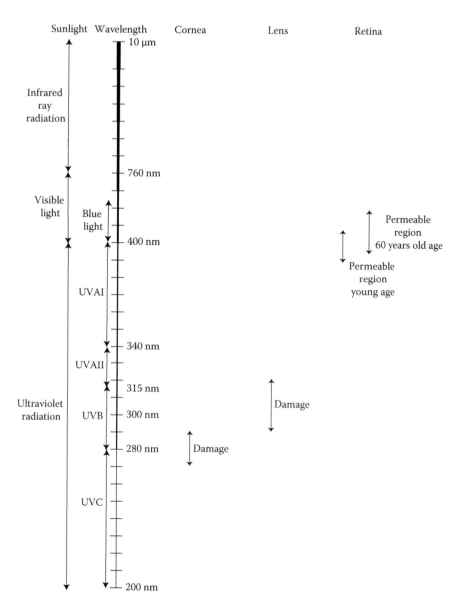

FIGURE 5.15
Sunlight and beneficial versus harmful effects on the eye.

Conjunctival or uveal melanocytes are exposed to UV and visible radiations, and the eye is subject to the development of ocular melanoma (Hu, 2005).

The retina is rich in visual pigment chromophores such as 11-*cis* retinal, and 11-*cis* retinal–protein complexes absorb visible light of all wavelengths.

Rods and cones participate in light selectivity, where melanin and lipofuscin exhibit broadband absorbance of low-wavelength energy and are associated with damage following exposure to blue light. Hemoglobin and most proteins absorb wavelengths between 400 and 600 nm or less; flavins and flavoproteins absorb blue light at 450 nm; and macular pigment absorbs wavelengths of 400–530 nm.

High-level light exposure induces moderate thermal damage to the retina that can present as irreversible thermal injury. Prolonged exposure to light can lead to cell death via thermal and photochemical damage. Thermal injury results from the actions of visible and near-infrared radiations, while photochemical damage is caused by exposure to visible and UV radiations.

Visible light consists of wavelengths of ≤600 nm and comprises blue light (400–450 nm) (Roberts, 2011) and green light (490–580 nm) (Morgan et al., 2008), as well as violet, indigo, yellow, orange, and red light. Blue light mainly affects the retinal pigment epithelium, while certain green light wavelengths affect photoreceptor cells and the retinal pigment epithelium. Morgan et al. (2008) studied retinal damage in macaque monkeys exposed to green or infrared light of 2–150-fold normal intensity. The retina was observed by an adaptive optics scanning laser ophthalmoscope at 15 min after exposure to 568 (green) or 830 nm (infrared) light. Although a moderate decrease in autofluorescence was observed in the retinal pigment epithelium after exposure to light at 568 (but not 830) nm, the decrease was reversible. The reversibility of autofluorescence and long-term damage to the retinal pigment epithelium were discerned in a long-term observation period of 5–165 days. Damage after long-term exposure to green light at 568 nm was reported as a safety concern by the U.S. National Standard Agency.

5.4.3 Photostability of Nanoproducts

Nanoparticles reportedly demonstrate enhanced photostability relative to other nanomaterials (Al-Sherbini, 2010). Photostability of gold nanorods (5 × 50 nm) and gold nanoparticles (micelle formulation) was examined following 20 min exposure to light at 400–900 nm (Al-Sherbini, 2010). The optical density of the gold nanorod decreased with each 20 min of exposure, whereas the gold nanoparticles in the micelle showed no substantial changes. TEM images indicated that the unencapsulated gold nanorods were degraded and modified into irregular shapes. In another study, a nanoemulsion of 290–500 nm particles loaded with resveratrol was administered to porcine skin on the ear (Carlotti et al., 2012). The emulsion formulation improved the stability of resveratrol, as assessed by preservation of its antilipoperoxidase activity.

5.4.4 Safety of Nanoparticles Delivered via the Integumentary System

Surface coatings contribute to the safety of metallic nanomaterials (Johnston et al., 2009). Although uncoated metallic nanomaterials (e.g., titanium dioxide or zinc oxide nanoparticles) can induce free radical effects (Dunford et al., 1997; Nakagawa et al., 1997; Wamer et al., 1997; Hirakawa et al., 2004; Gurr et al., 2005; Pinto et al., 2010), this tendency is attenuated by surface coating (Wakefield et al., 2004; Pan et al., 2009). An in vitro study of coated and uncoated 15 nm rutile and 200 nm anatase titanium dioxide nanoparticles demonstrated a negative influence of uncoated nanoparticles on skin fibroblast cell number, cell migration, and traction stress. Uptake analysis revealed large-scale uptake after 2 days of incubation with uncoated nanoparticles (0.4 mg/mL). However, the coated nanoparticles did not alter cell number even after 11 days of incubation at a concentration of 0.8 mg/mL, suggesting the cytoprotective effect of coating (Pan et al., 2009).

Nanoparticles do not easily penetrate into deep regions of the skin. Nanoparticle permeance was explored for three kinds of titanium dioxide nanoparticles: uncoated 300–500 nm nanoparticles (group A), uncoated 30–50 nm nanoparticles (group B), and coated nanorods ($20–30 \times 60–150$ nm; group C) (Sadrieh et al., 2010). The nanoparticles were administered to four regions of porcine skin (neck, back, abdomen, and tail regions) for 5 days/week for 4 weeks (total dose $= 176$ mg/cm^2). The contents of titanium in the epidermal and dermal layers and the presence of nanoparticles in the epidermis were monitored by TEM on the day following administration.

Titanium contents in the untreated control and groups A, B, and C averaged to 0.035, 2.11, 2.57, and 5.05 mg/g, respectively, in the epidermis and 1.003, 7.34, 5.17, and 10.92 µg/g, respectively, in the dermis. These values yielded epidermis-to-dermis ratios of 35:1, 287:1, 497:1, and 462:1, respectively. TEM showed a low frequency of nanoparticle presence in the epidermis, but when they were detected, the three nanoproducts were located at the surface of the keratinocyte layer and the ventral region of the upper follicular lumens. Therefore, the permeability of the skin, and especially the dermal region, was extremely low for all three titanium dioxide nanoproducts, regardless of particle size. No clear pathological conditions were observed in the skin in association with the nanoproducts.

5.4.5 Phototoxicity of Nanoparticles Administered via the Ocular Route

Phototoxicity of substances delivered via the ocular route has been studied by using water-soluble, hydroxylated fullerenes and lens epithelial cells (Roberts et al., 2008; Wielgus et al., 2010). Fullerenes are used as drug carriers to bypass the blood–optical barrier. α-Crystalline, a water-soluble structural protein, is the major lens protein found in human lens epithelial cells and comprises up to 40% of all nuclear lens proteins. Hydroxylated fullerenes exist as particles with three different sizes (21, 374, and 5054 nm) upon nonspecific binding to human serum albumin and two different sizes (23 and 365 nm) upon nonspecific binding to α-crystalline in lens epithelial cells. The average fullerene size at neutral pH is 659 nm due to agglomeration by van der Waals forces (Roberts et al., 2008).

The in vitro phototoxicity of hydroxylated fullerene nanoparticles was examined for UVA and visible light ranges, including blue and green light (Roberts et al., 2008). Accumulation of fullerenes in HLE B-3 human lens epithelial cells was confirmed after exposure to blue light (405 nm). Fullerene nanoparticles were cytotoxic to HLE B-3 cells maintained in the dark at concentrations higher than 20 μM, but they demonstrated increased cytotoxicity after cellular exposure to UVA or visible radiation at fullerene concentrations of ≥5 μM. Cytotoxicity upon UVA exposure was more remarkable than that for visible light exposure.

Fullerene nanoparticle cytotoxicity occurred as a result of early apoptotic rather than necrotic cell death. Early apoptosis is indicative of membrane damage to lens cells, where membrane damage can precipitate the development of cataracts. The authors concluded that the fullerene nanoparticles stimulated the generation of reactive oxygen intermediates because their cytotoxicity at 15 μM following exposure to UVA rays was ameliorated by cotreatment with the endogenous antioxidant lutein (20 μM) (Roberts et al., 2008).

Phototoxicity and in vitro cytotoxicity of water-soluble, hydroxylated fullerene nanoparticles were examined in another study of retinal pigment epithelial cells (Wielgus et al., 2010). Fullerene nanoparticles (1–50 μM) with a diameter of 364 nm were administered to the cells over a period of 17 h. Fullerene nanoparticle cytotoxicity was observed following cell exposure to visible radiation (≥400 nm) at 10–50 μM. However, cell death was not induced in the dark at even an extremely high fullerene concentration of 50 M. Free radicals were produced at 50 μM fullerene in the dark and at 10 μM or more in the light.

5.5 Factors Affecting Permeability and Disposition in the Integumentary System

The delivery of nanoproducts via the integumentary system includes three pathways: dermal, follicular, and sweat glands (Figure 5.16) (Barry, 2002; Benson, 2005). The follicular densities in the skin of the lateral forehead, thorax, back, upper arm, forearm, thigh, and calf in humans are 292, 22, 29, 32, 18, 17, and 14 follicles/cm^2, respectively. The follicular density of the lateral forehead is thus 10-fold higher or more than the density of the other follicular regions (Otberg et al., 2004).

With the exception of head hair, the follicular area and density in the human integumentary system are decreased compared with the area and density in most animals, such as rats or mice. The extrapolation to humans from studies of animals with abundant hair should therefore be adjusted accordingly (Howes et al., 1996), if the dermal permeability is not assumed

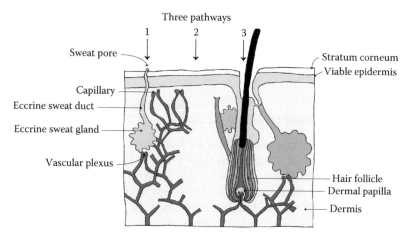

FIGURE 5.16

Three pathways of integumentary delivery: (1) sweat gland, (2) dermal route, and (3) follicular route. (Modified from Barry, B.W., *Adv. Drug Deliv. Rev.*, 54, S31, 2002; Benson, H.A.E., *Curr. Drug Deliv.*, 2(1), 23, 2005, doi: 10.2174/1567201052772915; Cevc, G. and Vierl, U., *J. Control. Rel.*, 118(1), 18, 2007, doi: 10.1016/j.jconrel.2006.10.022.)

to be the same as follicular permeability. Indeed, investigations of functionalized fullerenes (Xia et al., 2010), quantum dots (Ryman-Rasmussen et al., 2006, 2007), and maghemite (Baroli et al., 2007) penetration into the skin showed that nanoparticles of ≤1–10 nm passed through the keratinocyte layer, while nanoparticles of ≥10 nm can pass through the hair follicles but not the keratinized layer.

5.5.1 Transdermal Route

Pharmacokinetics of nanoproducts delivered via the transdermal route are affected by (1) the thickness of the keratinocyte layer, (2) the hydrophilic and hydrophobic pores in the keratinocyte layer, (3) the stimulation of the keratinocyte layer, and (4) the physicochemical parameters of nanoproducts (Potts and Guy, 1992; Lipinski et al., 2001; Choy and Prausnitz, 2011). I discuss each of these in the following detail.

5.5.1.1 Thickness of Keratinocyte Layer

The fluid volume of the keratinocyte layer of human skin is ~20%–30% under dry conditions (Nakazawa et al., 2012). The thickness of the keratinocyte layer changes by skin hydration level. A study of water distribution and dilation in the human keratinocyte layer showed that fluid volume can reach a midrange of 57%–87% in exogenously hydrated skin, which exceeds the natural skin hydration level of 30%–50% (Bouwstra et al., 2003). The hydration domain is mainly observed in the dead keratinocyte layer as opposed to the live cell layer. When skin is subjected to exogenous hydrating conditions such that the hydration level reaches 300%, the keratinocyte layer becomes sufficiently dilated that the hydration domain extends even to the live cell layer. Moreover, the thickness of the keratinocyte layer increases in a linear fashion at a hydration level of 17%–300%.

5.5.1.2 Hydrophilic and Hydrophobic Pores

A study of hydrophilic, water-soluble substances for skin modeling demonstrated that particles of ≤10 nm can pass through the pores in the keratinocyte layer (Mitragotri, 2003). Another study showed that the pores in the skin of hairless rats were 7.2 nm in diameter (Ruddy and Hadzija, 1992).

The permeability of the keratinocyte layer to lipid-soluble substances depends on the molecular weight and partition coefficient rather than the size of the substance in question (Potts and Guy, 1992; Mitragotri, 2002, 2003; Moss et al., 2002; Geinoz et al., 2004; Wang et al., 2006, 2007; Kretsos et al., 2008; Kushner et al., 2008; Lian et al., 2008). Lipid-soluble substances of ≥400–500 Da are trapped in the keratinocyte layer (Johnson et al., 1997; Mitragotri, 2003).

5.5.1.3 Stimulation of the Keratinocyte Layer

The permeability of the transdermal pathway is enhanced by strong electrical stimulation of keratinocytes, in addition to mechanical stimulation (e.g., sound or magnetism) and exposure to thermal stimuli (Cevc, 1996; Schatzlein and Cevc, 1998). For example, the dermal route allows passage of anionic, water-soluble substances of ~40 nm or less because these substances facilitate pore expansion by electroosmosis (Aguilella et al., 1994). The size of ionophoretically expanded pores is 15 nm in the rat (Manabe et al., 2000), and naturally occurring pores are 2–50-fold smaller than expanded pores (Cevc and Vierl, 2010). The recovery time to return to equilibrium after electrical stimulation depends on both membrane porosity and pore geometry. Furthermore, pore size in the skin of human cadavers is ~2.5 nm and increases with increasing salt concentration (Dinh et al., 1993), and pore size in porcine skin is ~1.4–3.4 nm following ionophoretic stimulation (Lai and Roberts, 1999).

5.5.1.4 Physicochemical Parameters of Nanoproducts

High-throughput screening enables the rapid assessment of an enormous number of candidate drugs by experimental and computational approaches. High-throughput screening was initially employed around year 1990 to estimate the solubility and permeability of candidate drugs during the experimental research and development stages. Computational methods in high-throughput screening utilize the rule-based Moriguchi Log P (MLogP) calculation (Moriguchi et al., 1992). The "Rule of 5" was proposed by Lipinski et al. (2001) for selecting candidates in the research phase from 2245 drugs by considering four physicochemical parameters from the U.S. Adopted Names database (A–D), with one exception (E) made for certain HMW substrates (peptides, some polymers):

A. There are more than five hydrogen bond donors (expressed as the sum of hydroxyl [OH] and amine [NH] groups).

B. The molecular weight is over 500 Da.

C. The Log P is over 5 (or the MLogP is over 4.15).

D. There are more than 10 hydrogen bond acceptors (expressed as the sum of nitrogen and oxygen atoms).

E. Compounds that are substrates for biological transporters are exceptions to the rule.

The probabilities of parameters A, B, C, and D are 8%, 11%, 10%, and 12%, respectively. Combinations of parameters change the probability of occurrence. For example, the probabilities of the combinations of A+D, D+B, A+B, and B+C are 10%, 7%, 4%, and 1%, respectively. These findings suggest that the "Rule of 5" is beneficial for selecting oral drug candidates and drugs for ophthalmic delivery, whereas the rule is not beneficial for selecting drug candidates for delivery via inhalation or the subcutaneous route (Choy and Prausnitz, 2011). The application of the "Rule of 5" is also limited in regard to drug administration via inhalation because the dermal epithelium has a thicker keratinocyte layer than the epithelium of the respiratory system.

The "Rule of 5" is also beneficial when nanoparticles are degraded and active ingredients are released. However, the estimation by the "Rule of 5" may not be accurate for intact nanoparticles in normal skin without keratinocyte stimulation.

5.5.2 Follicular Route

The pharmacokinetics of nanoproducts delivered via the pilosebaceous unit depend on the size selectivity of the pilosebaceous orifice and the presence of sebum, the hair growth cycle, stimulation by massage, and the physicochemical properties of the administered nanoproducts (Meidan et al., 2005; Wosicka and Cal, 2010). The orifices of the sweat gland, follicle, and sebaceous gland allow entry and egress of water-soluble substances and are ~50, 5–70, and 5–15 μm in diameter, respectively (Mitragotri, 2003). Nanoscale substances exposed to the integumentary system frequently distribute to the hair follicle (Betz et al., 2001; Alvarez-Roman et al., 2004; Jung et al., 2006).

The permeability of the follicular unit gland was examined by using the moderately lipid-soluble compound curcumin (Lademann et al., 2001). Curcumin passed through actively growing follicular cells, but not through resting follicular cells. Furthermore, DNA transfer into follicular cells occurred during the early phases of the growth cycle (Domashenko et al., 2000), while the distribution of curcumin was limited to the growth phase (Lademann et al., 2001). Another study showed that the use of wetting agent enhanced drug contact with the sebaceous gland via the pore in the duct (Illel, 1997).

The effect of massage on pilosebaceous permeability was illustrated in a study showing that massage enhanced the dermal penetration and accumulation of 320 nm fluorescent nanoparticles in both pig and human (Lademann et al., 2009). The keratinocyte cuticle is ~500–800 nm thick according to a structural analysis of the hair surface and the hair follicle (Biel et al., 2003). Therefore, nanoparticle accumulation might be attributed to particle deposition in the space vacated by keratinocytes forcefully removed during massage.

The permeability of the follicle in the porcine ear was examined by using poly(lactic-co-glycolic acid) (PLGA) and silica nanoparticles (Patzelt et al., 2011). PLGA nanoparticles of 122, 230, 300, 470, 643, and 860 nm and silica nanoparticles of 300, 646, 920, and 1000 nm were utilized in the study. PLGA and silica nanoparticles of 400–700 nm crossed into the follicle, while only the 643 nm PLGA and 646 nm silica nanoparticles penetrated into the deepest portion of the follicular root. The smaller (300 nm silica, 122/230/300 nm PLGA) and larger particles (860 nm PLGA nanoparticle, 920 and 1000 nm silica nanoparticles) showed significantly lower penetration depths. While small and large particles (122 and 860 nm PLGA and 920 and 1000 nm silica) can be utilized for targeting the infundibular region, larger particles (230 and 300 nm PLGA and 300 nm silica) are qualified for penetrating into the region of the sebaceous gland (Figure 5.17a). The 470 and 643 nm particles penetrated deepest into the hair follicle down to the bulge region (Figure 5.17a). The permeability of the follicle in human skin was similarly examined by using polystyrene nano- and microspheres (Toll et al., 2004). Nanoparticles of 750 nm and microparticles of 1.5, 3.0, and 6.0 μm crossed into the deepest region of the follicle (Figure 5.17d), demonstrating reasonably good consistency in particle-size selectivity between pigs and humans.

The distribution and uptake of nanoparticles after exposure to porcine skin was further examined by using two fluorescent, nonbiodegradable polystyrene nanoparticles of 20 and 200 nm (Alvarez-Roman et al., 2004). Nanoparticles were observed in the epidermal skin furrows and the follicle at 30 min after administration and preferentially distributed into the follicle after 1 and 2 h. Accumulation was observed for both 20 and 200 nm particles but was most pronounced for the 20 nm nanoparticle (Figure 5.17b). However, the nanoparticles did not cross the epidermis, and no evidence was found for systemic uptake via the follicle.

Skin tissue removed from the human abdomen was used to examine the effects of short- (30 min) and long-term (144 h) exposure on the distribution of heparin-coated and uncoated anionic liposomes loaded with fluorescent dye (Betz et al., 2001). The heparin-coated liposomes distributed to

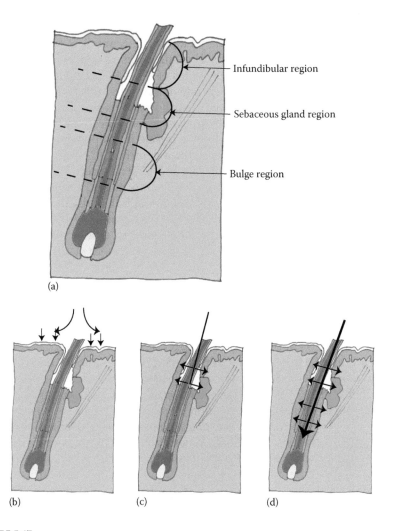

FIGURE 5.17
Transdermal and transfollicular routes under normal conditions without keratinocyte stimulation. (a) Penetration depths in relation to the target sites within hair follicles; (b) distribution to the orifice of the follicular unit; (c) distribution to the epidermis and dermis; and (d) size-dependent dermal penetration. (Modified from Vogt, A. et al., *J. Invest. Dermatol.*, 126(6), 1316, 2006, doi: 10.1038/sj.jid.5700226; Kajimoto, K. et al., *Int. J. Pharmaceut.*, 403(1–2), 57, 2011, doi: 10.1016/j.ijpharm.2010.10.021; Patzelt, A. et al., *J. Control. Rel.*, 150(1), 45, 2011, doi: 10.1016/j.jconrel.2010.11.015.)

deep dermal regions regardless of exposure time or particle size. On the other hand, uncoated 200–693 nm liposomes distributed to a depth of 5 μm, corresponding to a location in the epidermal layer, while heparin-coated 315–743 nm liposomes distributed to a depth of ≥10 μm and specifically to the hair shaft, corresponding to a location deep in the dermis (Figure 5.17c).

The distribution of five different liposomes after exposure to hair follicles in the porcine ear was examined by biopsy at 1 h and 3, 5, and 7 days after administration (Jung et al., 2006). The liposomes varied by size, lipid constitution, and surface charge. Liposomes of 231–448 nm distributed to deep regions of the skin in a size-independent, but polar and charge-dependent, manner. The liposomes all showed distribution to a similar depth after 1 h and 3 days. After 5 and 7 days, 231 and 234 nm anionic liposomes distributed to a depth of 40%, while 319 nm polar but neutral liposomes and 448 and 253 nm cationic liposomes distributed to a depth of 50%–70% (Figure 5.17d).

Three fluorescent particles of 40, 750, and 1500 nm were transcutaneously applied to human skin (Vogt et al., 2006). Only the 40 nm particle deeply penetrated into vellus hair openings and through the follicular epithelium.

Transfollicular delivery by ionophoresis was examined by using anionic or cationic liposomes loaded with rhodamine to assess pharmacokinetic parameters or with insulin to assess efficacy (Kajimoto et al., 2011). Anionic liposomes failed to transfer into the skin. Cationic, rhodamine-loaded, 288 nm liposomes with an equilibrium potential of +61 mV were transferred into the skin via transfollicular units in normal rats, while cationic, insulin-loaded, 243 nm liposomes with an equilibrium potential of +63 mV showed a continuous and gradual decrease in blood glucose over an 18 h period after administration to streptozotocin-induced diabetic model rats.

Penetration of 20 nm titanium dioxide nanoparticles into the hair follicles of both human and porcine skin was examined by microscopy and autoradiography (Lekki et al., 2007). Primary 20 nm particles traveled as far as ~400 μm into the follicle, without accumulation in vital tissue or the sebaceous gland.

Delivery via follicular antigen-presenting cells is attributed to M cells in the GI tract (Mahe et al., 2009). Penetration into the skin and protection against viruses were examined in mice by using fluorescent 40 or 200 nm polystyrene nanoparticles, as well as fluorescent 290 nm virus-derived nanoparticles loaded with an antiviral vaccine (Mahe et al., 2009). The 40 nm nanoparticles penetrated to a greater depth (80 μm) than the 200 and 290 nm nanoparticles, followed by entry into hair follicles and internalization by perifollicular antigen-presenting cells (Figure 5.17d). However, the 290 nm virus nanoparticle protected against viral infection.

5.6 Nanoscale World via the Integumentary System and the Ocular Route

The sizes of nanoproducts, tight junctions, and anatomical pores related to the integumentary system and the eye are described in this chapter and summarized in Figure 5.18. The tight junction space of the dermal keratinocyte is ~45 nm, while the keratinocyte pore affords a permeability barrier of 5 nm or

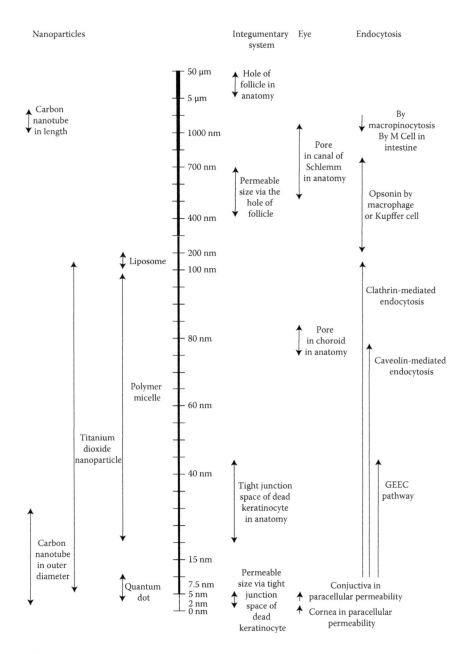

FIGURE 5.18
The size of nanoproducts administered via the integumentary system and the eye, and the size of relevant anatomical features and permeability barriers.

less in humans. The orifice of the hair follicle within the pilosebaceous unit is observed as 5–70 µm in anatomy, while in terms of its permeability, it is actually only ~400–700 nm.

The sizes of the ocular pores in the canal of Schlemm range from 500 nm to 1.5 µm. Nanoparticles with a 500 nm diameter pass through the outflow pathway with a frequency of 50%, showing a similar distribution as conventional drugs. The size of the choroid pore is ~75–85 nm in rodents. The paracellular space of the corneal epithelium is ~2 nm, while that of the palpebral conjunctiva is ~4.9 nm. Therefore, these pathways are too narrow to permit passage of nanoscale materials.

References

Aguilella, V., K. Kontturi, L. Murtomaki, and P. Ramirez. 1994. Estimation of the pore-size and charge-density in human cadaver skin. *Journal of Controlled Release* 32 (3):249–257. doi: 10.1016/0168-3659(94)90235-6.

Ahamed, I. and T. F. Patton. 1985. Importance of the noncorneal absorption route in topical ophthalmic drug delivery. *Investigative Ophthalmology & Visual Science* 26:584–587.

Al-Amoudi, A., J. Dubochet, and L. Norlen. 2005. Nanostructure of the epidermal extracellular space as observed by cryo-electron microscopy of vitreous sections of human skin. *Journal of Investigative Dermatology* 124 (4):764–777. doi: 10.1111/j.0022-202X.2005.23630.x.

Ali, Y. and K. Lehmussaari. 2006. Industrial perspective in ocular drug delivery. *Advanced Drug Delivery Reviews* 58 (11):1258–1268. doi: 10.1016/j.addr.2006.07.022.

Alonso, J. L. and W. H. Goldmann. 2003. Feeling the forces: Atomic force microscopy in cell biology. *Life Sciences* 72 (23):2553–2560. doi: 10.1016/S0024-3205(03)00165-6.

Al-Sherbini, E. S. A. M. 2010. UV-visible light reshaping of gold nanorods. *Materials Chemistry and Physics* 121 (1–2):349–353. doi: 10.1016/j.matchemphys.2010.01.048.

Alvarez-Roman, R., A. Naik, Y. Kalia, R. H. Guy, and H. Fessi. 2004. Skin penetration and distribution of polymeric nanoparticles. *Journal of Controlled Release* 99 (1):53–62. doi: 10.1016/j.jconrel.2004.06.015.

Amrite, A. C. and U. B. Kompella. 2005. Size-dependent disposition of nanoparticles and microparticles following subconjunctival administration. *Journal of Pharmacy and Pharmacology* 57 (12):1555–1563. doi: 10.1211/jpp.57.12.0005.

Andrady, A. L., P. J. Aucamp, A. T. Austin, A. F. Bais, C. L. Ballare, L. O. Bjorn, J. F. Bornman et al., and Panel Environmental Effects Assessment. 2012. Environmental effects of ozone depletion and its interactions with climate change: Progress report, 2011 United Nations Environment Programme, Environmental Effects Assessment Panel. *Photochemical & Photobiological Sciences* 11 (1):13–27. doi: 10.1039/c1pp90033a.

Anunciato, T. P. and P. A. da Rocha. 2012. Carotenoids and polyphenols in nutricosmetics, nutraceuticals, and cosmeceuticals. *Journal of Cosmetic Dermatology* 11 (1):51–54. doi: 10.1111/j.1473-2165.2011.00600.x.

Araujo, R., M. Fernandes, A. Cavaco-Paulo, and A. Gomes. 2011. Biology of human hair: Know your hair to control it. *Biofunctionalization of Polymers and Their Applications* 125:121–143. doi: 10.1007/10_2010_88.

Baroli, B., M. G. Ennas, F. Loffredo, M. Isola, R. Pinna, and M. A. Lopez-Quintela. 2007. Penetration of metallic nanoparticles in human full-thickness skin. *Journal of Investigative Dermatology* 127 (7):1701–1712. doi: 10.1038/sj.jid.5700733.

Barry, B. W. 2002. Drug delivery routes in skin: A novel approach. *Advanced Drug Delivery Reviews* 54:S31–S40.

Beissert, S. and R. D. Granstein. 1996. UV-induced cutaneous photobiology. *Critical Reviews in Biochemistry and Molecular Biology* 31 (5–6):381–404.

Benson, H. A. E. 2005. Transdermal drug delivery: Penetration enhancement techniques. *Current Drug Delivery* 2 (1):23–33. doi: 10.2174/1567201052772915.

Betz, G., R. Imboden, and G. Imanidis. 2001. Interaction of liposome formulations with human skin in vitro. *International Journal of Pharmaceutics* 229 (1–2):117–129.

Biel, S. S., K. Kawaschinski, K. P. Wittern, U. Hintze, and R. Wepf. 2003. From tissue to cellular ultrastructure: Closing the gap between micro- and nanostructural imaging. *Journal of Microscopy (Oxford)* 212:91–99. doi: 10.1046/j.1365-2818.2003.01227.x.

Birbeck, M. S. C. and E. H. Mercer. 1957a. The electron microscopy of the human hair follicle. 1. Introduction and the hair cortex. *Journal of Biophysical and Biochemical Cytology* 3 (2):203–214.

Birbeck, M. S. C. and E. H. Mercer. 1957b. The electron microscopy of the human hair follicle. 2. The hair cuticle. *Journal of Biophysical and Biochemical Cytology* 3 (2):215–221.

Birbeck, M. S. C. and E. H. Mercer. 1957c. The electron microscopy of the human hair follicle. 3. The inner root sheath and trichohyaline. *Journal of Biophysical and Biochemical Cytology* 3 (2):223–229.

Boncheva, M., J. de Sterke, P. J. Caspers, and G. J. Puppels. 2009. Depth profiling of Stratum corneum hydration in vivo: A comparison between conductance and confocal Raman spectroscopic measurements. *Experimental Dermatology* 18 (10):870–876. doi: 10.1111/j.1600-0625.2009.00868.x.

Boulton, M., M. Rozanowska, and B. Rozanowski. 2001. Retinal photodamage. *Journal of Photochemistry and Photobiology B—Biology* 64 (2–3):144–161.

Bouwstra, J. A., A. de Graaff, G. S. Gooris, J. Nijsse, J. W. Wiechers, and A. C. van Aelst. 2003. Water distribution and related morphology in human stratum corneum at different hydration levels. *Journal of Investigative Dermatology* 120 (5):750–758.

Braverman, I. M. 1997. The cutaneous microcirculation: Ultrastructure and microanatomical organization. *Microcirculation (London)* 4 (3):329–340. doi: 10.3109/10739689709146797.

Bressler, N. M. and S. B. Bressler. 2000. Photodynamic therapy with verteporfin (visudyne): Impact on ophthalmology and visual sciences. *Investigative Ophthalmology & Visual Science* 41 (3):624–628.

Carlotti, M. E., S. Sapino, E. Ugazio, M. Gallarate, and S. Morel. 2012. Resveratrol in solid lipid nanoparticles. *Journal of Dispersion Science and Technology* 33 (4): 465–471. doi: 10.1080/01932691.2010.548274.

Caspers, P. J., G. W. Lucassen, H. A. Bruining, and G. J. Puppels. 2000. Automated depth-scanning confocal Raman microspectrometer for rapid in vivo determination of water concentration profiles in human skin. *Journal of Raman Spectroscopy* 31 (8–9):813–818. doi: 10.1002/1097-4555(200008/09)31:8/9<813::aid-jrs573>3.0.co;2-7.

Cevc, G. 1996. Transfersomes, liposomes and other lipid suspensions on the skin: Permeation enhancement, vesicle penetration, and transdermal drug delivery. *Critical Reviews in Therapeutic Drug Carrier Systems* 13 (3–4):257–388.

Cevc, G. and U. Vierl. 2007. Spatial distribution of cutaneous microvasculature and local drug clearance after drug application on the skin. *Journal of Controlled Release* 118 (1):18–26. doi: 10.1016/j.jconrel.2006.10.022.

Cevc, G. and U. Vierl. 2010. Nanotechnology and the transdermal route A state of the art review and critical appraisal. *Journal of Controlled Release* 141 (3):277–299. doi: 10.1016/j.jconrel.2009.10.016.

Chen, T. C., F. Chimeh, Z. R. Lu, J. Mathieu, K. S. Person, A. Q. Zhang, N. Kohn, S. Martinello, R. Berkowitz, and M. F. Holick. 2007. Factors that influence the cutaneous synthesis and dietary sources of vitamin D. *Archives of Biochemistry and Biophysics* 460 (2):213–217. doi: 10.1016/j.abb.2006.12.017.

Choy, Y. B. and M. R. Prausnitz. 2011. The rule of five for non-oral routes of drug delivery: Ophthalmic, inhalation and transdermal. *Pharmaceutical Research* 28 (5):943–948. doi: 10.1007/s11095-010-0292-6.

Couteau, C., E. Paparis, S. El-Bourry-Alami, and L. J. M. Coiffard. 2012. Influence on SPF of the quantity of sunscreen product applied. *International Journal of Pharmaceutics* 437 (1–2):250–252. doi: 10.1016/j.ijpharm.2012.08.019.

de Gruijl, F. R., H. J. C. M. Sterenborg, P. D. Forbes, R. E. Davies, C. Cole, G. Kelfkens, H. Vanweelden, H. Slaper, and J. C. Vanderleun. 1993. Wavelength dependence of skin-cancer induction by ultraviolet-irradiation of albino hairless mice. *Cancer Research* 53 (1):53–60.

Diebold, Y. and M. Calonge. 2010. Applications of nanoparticles in ophthalmology. *Progress in Retinal and Eye Research* 29 (6):596–609. doi: 10.1016/j.preteyeres.2010.08.002.

Dillon, J. 1991. The photophysics and photobiology of the eye. *Journal of Photochemistry and Photobiology B—Biology* 10 (1–2):23–40. doi: 10.1016/1011-1344(91)80209-z.

Dinh, S. M., C. W. Luo, and B. Berner. 1993. Upper and lower limits of human skin electrical-resistance in iontophoresis. *AIChE Journal* 39 (12):2011–2018. doi: 10.1002/aic.690391211.

Domashenko, A., S. Gupta, and G. Cotsarelis. 2000. Efficient delivery of transgenes to human hair follicle progenitor cells using topical lipoplex. *Nature Biotechnology* 18 (4):420–423.

Draelos, Z. D. 2010. Nutrition and enhancing youthful-appearing skin. *Clinics in Dermatology* 28 (4):400–408. doi: 10.1016/j.clindermatol.2010.03.019.

Dunford, R., E. J. Land, M. Rozanowska, T. Sarna, and T. G. Truscott. 1995. Interaction of melanin with carbon-centered and oxygen-centered radicals from methanol and ethanol. *Free Radical Biology and Medicine* 19 (6):735–740. doi: 10.1016/0891-5849(95)00059-7.

Dunford, R., A. Salinaro, L. Z. Cai, N. Serpone, S. Horikoshi, H. Hidaka, and J. Knowland. 1997. Chemical oxidation and DNA damage catalysed by inorganic sunscreen ingredients. *FEBS Letters* 418 (1–2):87–90. doi: 10.1016/s0014-5793(97)01356-2.

Fartasch, M., I. D. Bassukas, and T. L. Diepgen. 1993. Structural relationship between epidermal lipid lamellae, lamellar bodies and desmosomes in human epidermis—An ultrastructural study. *British Journal of Dermatology* 128 (1):1–9. doi: 10.1111/j.1365-2133.1993.tb00138.x.

Frennesson, C. I. and S. E. G. Nilsson. 2004. Encouraging results of photodynamic therapy with visudyne in a clinical patient material of age-related macular degeneration. *Acta Ophthalmologica Scandinavica* 82 (6):645–650.

Fujita, H. and T. Fujita. 1992. *Textbook of Histology: Part 2*, 3rd edn. Tokyo, Japan: Igaku-Shoin, Ltd.

Gaudana, R., J. Jwala, S. H. S. Boddu, and A. K. Mitra. 2009. Recent perspectives in ocular drug delivery. *Pharmaceutical Research* 26 (5):1197–1216. doi: 10.1007/s11095-008-9694-0.

Geinoz, S., R. H. Guy, B. Testa, and P. A. Carrupt. 2004. Quantitative structure-permeation relationships (QSPeRs) to predict skin permeation: A critical evaluation. *Pharmaceutical Research* 21 (1):83–92. doi: 10.1023/B:PHAM.0000012155.27488.2b.

Ghate, D., W. Brooks, B. E. McCarey, and H. F. Edelhauser. 2007. Pharmacokinetics of intraocular drug delivery by periocular injections using ocular fluorophotometry. *Investigative Ophthalmology & Visual Science* 48 (5):2230–2237. doi: 10.1167/iovs.06-0954.

Gillbro, J. M. and M. J. Olsson. 2011. The melanogenesis and mechanisms of skin-lightening agents—Existing and new approaches. *International Journal of Cosmetic Science* 33 (3):210–221. doi: 10.1111/j.1468-2494.2010.00616.x.

Gulsen, D. and A. Chauhan. 2005. Dispersion of microemulsion drops in HEMA hydrogel: A potential ophthalmic drug delivery vehicle. *International Journal of Pharmaceutics* 292 (1–2):95–117. doi: 10.1016/j.ijpharm.2004.11.033.

Gurr, J. R., A. S. S. Wang, C. H. Chen, and K. Y. Jan. 2005. Ultrafine titanium dioxide particles in the absence of photoactivation can induce oxidative damage to human bronchial epithelial cells. *Toxicology* 213 (1–2):66–73. doi: 10.1016/j.tox.2005.05.007.

Guymer, R. H., A. C. Bird, and G. S. Hageman. 2004. Cytoarchitecture of choroidal capillary endothelial cells. *Investigative Ophthalmology & Visual Science* 45 (6):1660–1666. doi: 10.1167/Iovs.03-0913.

Halliday, G. M., N. S. Agar, R. S. C. Barnetson, H. N. Ananthaswamy, and A. M. Jones. 2005. UV-A fingerprint mutations in human skin cancer. *Photochemistry and Photobiology* 81 (1):3–8.

Halliday, G. M., S. N. Byrne, and D. L. Damian. 2011. Ultraviolet A radiation: Its role in immunosuppression and carcinogenesis. *Seminars in Cutaneous Medicine and Surgery* 30 (4):214–221. doi: 10.1016/j.sder.2011.08.002.

Hamalainen, K. M., K. Kananen, S. Auriola, K. Kontturi, and A. Urtti. 1997. Characterization of paracellular and aqueous penetration routes in cornea, conjunctiva, and sclera. *Investigative Ophthalmology & Visual Science* 38 (3):627–634.

Hirakawa, K., M. Mori, M. Yoshida, S. Oikawa, and S. Kawanishi. 2004. Photo-irradiated titanium dioxide catalyzes site specific DNA damage via generation of hydrogen peroxide. *Free Radical Research* 38 (5):439–447. doi: 10.1080/1071576042000206487.

Hironaka, K., T. Fujisawa, H. Sasaki, Y. Tozuka, K. Tsuruma, M. Shimazawa, H. Hara, and H. Takeuchi. 2011. Fluorescence investigation of the retinal delivery of hydrophilic compounds via liposomal eyedrops. *Biological & Pharmaceutical Bulletin* 34 (6):894–897.

Hironaka, K., Y. Inokuchi, Y. Tozuka, M. Shimazawa, H. Hara, and H. Takeuchi. 2009. Design and evaluation of a liposomal delivery system targeting the posterior segment of the eye. *Journal of Controlled Release* 136 (3):247–253. doi: 10.1016/j.jconrel.2009.02.020.

Holbrook, K. A. and G. F. Odland. 1974. Regional differences in thickness (cell layers) of human stratum-corneum—Ultrastructural analysis. *Journal of Investigative Dermatology* 62 (4):415–422. doi: 10.1111/1523-1747.ep12701670.

Howes, D., R. Guy, J. Hadgraft, J. Heylings, U. Hoeck, F. Kemper, H. Maibach et al. 1996. Methods for assessing percutaneous absorption—The report and recommendations of ECVAM workshop 13. *ATLA—Alternatives to Laboratory Animals* 24 (1):81–106.

Hu, D. N. 2005. Photobiology of ocular melanocytes and melanoma. *Photochemistry and Photobiology* 81 (3):506–509.

Hunter, J. J., J. I. W. Morgan, W. H. Merigan, D. H. Sliney, J. R. Sparrow, and D. R. Williams. 2012. The susceptibility of the retina to photochemical damage from visible light. *Progress in Retinal and Eye Research* 31 (1):28–42. doi: 10.1016/j.preteyeres.2011.11.001.

Illel, B. 1997. Formulation for transfollicular drug administration: Some recent advances. *Critical Reviews in Therapeutic Drug Carrier Systems* 14 (3):207–219.

Inokuchi, Y., K. Hironaka, T. Fujisawa, Y. Tozuka, K. Tsuruma, M. Shimazawa, H. Takeuchi, and H. Hara. 2010. Physicochemical properties affecting retinal drug/coumarin-6 delivery from nanocarrier systems via eyedrop administration. *Investigative Ophthalmology & Visual Science* 51 (6):3162–3170. doi: 10.1167/Iovs.09-4697.

Ito, S. and K. Wakamatsu. 2003. Quantitative analysis of eumelanin and pheomelanin in humans, mice, and other animals: A comparative review. *Pigment Cell Research* 16 (5):523–531. doi: 10.1034/j.1600-0749.2003.00072.x.

Johnson, M. E., D. Blankschtein, and R. Langer. 1997. Evaluation of solute permeation through the stratum corneum: Lateral bilayer diffusion as the primary transport mechanism. *Journal of Pharmaceutical Sciences* 86 (10):1162–1172. doi: 10.1021/js960198e.

Johnson, M., D. Chan, A. T. Read, C. Christensen, A. Sit, and C. R. Ethier. 2002. The pore density in the inner wall endothelium of Schlemm's canal of glaucomatous eyes. *Investigative Ophthalmology & Visual Science* 43 (9):2950–2955.

Johnson, M., D. H. Johnson, R. D. Kamm, A. W. DeKater, and D. L. Epstein. 1990. The filtration characteristics of the aqueous outflow system. *Experimental Eye Research* 50 (4):407–418.

Johnston, H. J., G. R. Hutchison, F. M. Christensen, S. Peters, S. Hankin, and V. Stone. 2009. Identification of the mechanisms that drive the toxicity of TiO_2 particulates: The contribution of physicochemical characteristics. *Particle and Fibre Toxicology* 6:33. doi: 10.1186/1743-8977-6-33.

Jung, S., N. Otberg, G. Thiede, H. Richter, W. Sterry, S. Panzner, and J. Lademann. 2006. Innovative liposomes as a transfollicular drug delivery system: Penetration into porcine hair follicles. *Journal of Investigative Dermatology* 126 (8):1728–1732. doi: 10.1038/sj.jid.5700323.

Kajimoto, K., M. Yamamoto, M. Watanabe, K. Kigasawa, K. Kanamura, H. Harashima, and K. Kogure. 2011. Noninvasive and persistent transfollicular drug delivery system using a combination of liposomes and iontophoresis. *International Journal of Pharmaceutics* 403 (1–2):57–65. doi: 10.1016/j.ijpharm.2010.10.021.

Kaur, I. P. and M. Kanwar. 2002. Ocular preparations: The formulation approach. *Drug Development and Industrial Pharmacy* 28 (5):473–493.

Kealey, T., M. Philpott, and R. Guy. 1997. The regulatory biology of the human pilosebaceous unit. *Baillieres Clinical Obstetrics and Gynaecology* 11 (2):205–227.

Keister, J. C., E. R. Cooper, P. J. Missel, J. C. Lang, and D. F. Hager. 1991. Limits on optimizing ocular drug delivery. *Journal of Pharmaceutical Sciences* 80 (1):50–53. doi: 10.1002/jps.2600800113.

Kimlin, M. G. 2008. Geographic location and vitamin D synthesis. *Molecular Aspects of Medicine* 29 (6):453–461. doi: 10.1016/j.mam.2008.08.005.

Kleinpenning, M. M., T. Smits, M. H. A. Frunt, P. E. J. van Erp, P. C. M. van de Kerkhof, and R. M. J. P. Gerritsen. 2010. Clinical and histological effects of blue light on normal skin. *Photodermatology, Photoimmunology & Photomedicine* 26 (1):16–21.

Koeberle, M. J., P. M. Hughes, G. G. Skellern, and C. G. Wilson. 2006. Pharmacokinetics and disposition of memantine in the arterially perfused bovine eye. *Pharmaceutical Research* 23 (12):2781–2798. doi: 10.1007/s11095-006-9106-2.

Kourlas, H. and D. S. Schiller. 2006. Pegaptanib sodium for the treatment of neovascular age-related macular degeneration: A review. *Clinical Therapeutics* 28 (1):36–44. doi: 10.1016/j.clinthera.2006.01.009.

Kretsos, K., M. A. Miller, G. Zamora-Estrada, and G. B. Kasting. 2008. Partitioning, diffusivity and clearance of skin permeants in mammalian dermis. *International Journal of Pharmaceutics* 346 (1–2):64–79. doi: 10.1016/j.ijpharm.2007.06.020.

Kushner, J., D. Blankschtein, and R. Langer. 2008. Heterogeneity in skin treated with low-frequency ultrasound. *Journal of Pharmaceutical Sciences* 97 (10):4119–4128. doi: 10.1002/jps.21308.

Kuznetsova, T. G., M. N. Starodubtseva, N. I. Yegorenkov, S. A. Chizhik, and R. I. Zhdanov. 2007. Atomic force microscopy probing of cell elasticity. *Micron* 38 (8):824–833. doi: 10.1016/j.micron.2007.06.011.

Lademann, J., N. Otberg, H. Richter, H. J. Weigmann, U. Lindemann, H. Schaefer, and W. Sterry. 2001. Investigation of follicular penetration of topically applied substances. *Skin Pharmacology and Applied Skin Physiology* 14:17–22. doi: 10.1159/000056385.

Lademann, J., A. Patzelt, H. Richter, S. Schanzer, W. Sterry, A. Filbry, K. Bohnsack, F. Rippke, and M. Meinke. 2009. Comparison of two in vitro models for the analysis of follicular penetration and its prevention by barrier emulsions. *European Journal of Pharmaceutics and Biopharmaceutics* 72 (3):600–604. doi: 10.1016/j.ejpb.2009.02.003.

Lai, P. M. and M. S. Roberts. 1999. An analysis of solute structure-human epidermal transport relationships in epidermal iontophoresis using the ionic mobility: Pore model. *Journal of Controlled Release* 58 (3):323–333. doi: 10.1016/s0168-3659(98)00172-2.

Larsson, B. S. 1993. Interaction between chemicals and melanin. *Pigment Cell Research* 6 (3):127–133. doi: 10.1111/j.1600-0749.1993.tb00591.x.

Lekki, J., Z. Stachura, W. Dabros, J. Stachura, F. Menzel, T. Reinert, T. Butz et al. 2007. On the follicular pathway of percutaneous uptake of nanoparticles: Ion microscopy and autoradiography studies. *Nuclear Instruments & Methods in Physics Research, Section B—Beam Interactions with Materials and Atoms* 260 (1):174–177. doi: 10.1016/j.nimb.2007.02.021.

Lian, G. P., L. J. Chen, and L. J. Han. 2008. An evaluation of mathematical models for predicting skin permeability. *Journal of Pharmaceutical Sciences* 97 (1):584–598. doi: 10.1002/jps.21074.

Liebmann, J., M. Born, and V. Kolb-Bachofen. 2010. Blue-light irradiation regulates proliferation and differentiation in human skin cells. *Journal of Investigative Dermatology* 130 (1):259–269. doi: 10.1038/Jid.2009.194.

Lipinski, C. A., F. Lombardo, B. W. Dominy, and P. J. Feeney. 2001. Experimental and computational approaches to estimate solubility and permeability in drug discovery and development settings. *Advanced Drug Delivery Reviews* 46 (1–3):3–26. doi: 10.1016/s0169-409x(00)00129-0.

Liu, S. Y., L. Jones, and F. X. Gu. 2012. Nanomaterials for ocular drug delivery. *Macromolecular Bioscience* 12 (5):608–620. doi: 10.1002/mabi.201100419.

Mahe, B., A. Vogt, C. Liard, D. Duffy, V. Abadie, O. Bonduelle, A. Boissonnas, W. Sterry, B. Verrier, U. Blume-Peytavi, and B. Combadiere. 2009. Nanoparticle-based targeting of vaccine compounds to skin antigen-presenting cells by hair follicles and their transport in mice. *Journal of Investigative Dermatology* 129 (5):1156–1164. doi: 10.1038/Jid.2008.356.

Manabe, E., S. Numajiri, K. Sugibayashi, and Y. Morimoto. 2000. Analysis of skin permeation-enhancing mechanism of iontophoresis using hydrodynamic pore theory. *Journal of Controlled Release* 66 (2–3):149–158. doi: 10.1016/s0168-3659(99)00265-5.

Marmor, M. F., A. Negi, and D. M. Maurice. 1985. Kinetics of macromolecules injected into the subretinal space. *Experimental Eye Research* 40 (5):687–696. doi: 10.1016/0014-4835(85)90138-1.

Matousek, J. L. and K. L. Campbell. 2002. A comparative review of cutaneous pH. *Veterinary Dermatology* 13 (6):293–300. doi: 10.1046/j.1365-3164.2002.00312.x.

Matthews, Y. J., G. M. Halliday, T. A. Phan, and D. L. Damian. 2010. Wavelength dependency for UVA-induced suppression of recall immunity in humans. *Journal of Dermatological Science* 59 (3):192–197. doi: 10.1016/j.jdermsci.2010.07.005.

Meidan, V. M., M. C. Bonner, and B. B. Michniak. 2005. Transfollicular drug delivery—Is it a reality? *International Journal of Pharmaceutics* 306 (1–2):1–14. doi: 10.1016/j.ijpharm.2005.09.025.

Meredith, T. A. 2006. Intravitral antimicrobials. In: G. J. Jaffe, P. Ashton, and P. A. Pearson, eds. *Intraocular Drug Delivery*. New York: Taylor & Francis Group, pp. 85–95.

Mitchell, D. and A. Fernandez. 2012. The photobiology of melanocytes modulates the impact of UVA on sunlight-induced melanoma. *Photochemical & Photobiological Sciences* 11 (1):69–73. doi: 10.1039/C1pp05146f.

Mitragotri, S. 2002. A theoretical analysis of permeation of small hydrophobic solutes across the stratum corneum based on scaled particle theory. *Journal of Pharmaceutical Sciences* 91 (3):744–752. doi: 10.1002/jps.10048.

Mitragotri, S. 2003. Modeling skin permeability to hydrophilic and hydrophobic solutes based on four permeation pathways. *Journal of Controlled Release* 86 (1): 69–92. doi: 10.1016/s0168-3659(02)00321-8.

Moan, J., Z. Baturaite, A. C. Porojnicu, A. Dahlback, and A. Juzeniene. 2012. UVA, UVB and incidence of cutaneous malignant melanoma in Norway and Sweden. *Photochemical & Photobiological Sciences* 11 (1):191–198. doi: 10.1039/c1pp05215b.

Mohan, R. R., J. C. K. Tovey, A. Sharma, and A. Tandon. 2012. Gene therapy in the cornea: 2005—Present. *Progress in Retinal and Eye Research* 31 (1):43–64. doi: 10.1016/j.preteyeres.2011.09.001.

Morgan, J. I. W., J. J. Hunter, B. Masella, R. Wolfe, D. C. Gray, W. H. Merigan, F. C. Delori, and D. R. Williams. 2008. Light-induced retinal changes observed with high-resolution autofluorescence imaging of the retinal pigment epithelium. *Investigative Ophthalmology & Visual Science* 49 (8):3715–3729. doi: 10.1167/iovs.07-1430.

Moriguchi, I., S. Hirono, Q. Liu, I. Nakagome, and Y. Matsushita. 1992. Simple method of calculating octanol water partition-coefficient. *Chemical & Pharmaceutical Bulletin* 40 (1):127–130.

Moss, G. P., J. C. Dearden, H. Patel, and M. T. D. Cronin. 2002. Quantitative structure-permeability relationships (QSPRs) for percutaneous absorption. *Toxicology In Vitro* 16 (3):299–317. doi: 10.1016/s0887-2333(02)00003-6.

Mousa, S. A., D. J. Bharali, and D. Armstrong. 2007. From nutraceuticals to pharmaceuticals to nanopharmaceuticals: A case study in angiogenesis modulation during oxidative stress. *Molecular Biotechnology* 37 (1):72–80. doi: 10.1007/s12033-007-0064-7.

Mulholland, W. J., E. A. H. Arbuthnott, B. J. Bellhouse, J. F. Cornhill, J. M. Austyn, M. A. F. Kendall, Z. F. Cui, and U. K. Tirlapur. 2006. Multiphoton high-resolution 3D imaging of Langerhans cells and keratinocytes in the mouse skin model adopted for epidermal powdered immunization. *Journal of Investigative Dermatology* 126 (7):1541–1548. doi: 10.1038/sj.jid.5700290.

Nagel, E., W. Vilser, and I. Lanzl. 2003. Online human conjunctival vessel diameter analysis. A clinical-methodical study. *Clinical Hemorheology and Microcirculation* 28 (4):221–227.

Nakagawa, Y., S. Wakuri, K. Sakamoto, and N. Tanaka. 1997. The photogenotoxicity of titanium dioxide particles. *Mutation Research—Genetic Toxicology and Environmental Mutagenesis* 394 (1–3):125–132. doi: 10.1016/s1383-5718(97)00126-5.

Nakano, K., Y. Tozuka, H. Yamamoto, Y. Kawashima, and H. Takeuchi. 2008. A novel method for measuring rigidity of submicron-size liposomes with atomic force microscopy. *International Journal of Pharmaceutics* 355 (1–2):203–209. doi: 10.1016/j.ijpharm.2007.12.018.

Nakazawa, H., N. Ohta, and I. Hatta. 2012. A possible regulation mechanism of water content in human stratum corneum via intercellular lipid matrix. *Chemistry and Physics of Lipids* 165 (2):238–243. doi: 10.1016/j.chemphyslip.2012.01.002.

Otberg, N., H. Richter, H. Schaefer, U. Blume-Peytavi, W. Sterry, and J. Lademann. 2004. Variations of hair follicle size and distribution in different body sites. *Journal of Investigative Dermatology* 122 (1):14–19. doi: 10.1046/j.0022-202X.2003.22110.x.

Pan, Z., W. Lee, L. Slutsky, R. A. F. Clark, N. Pernodet, and M. H. Rafailovich. 2009. Adverse effects of titanium dioxide nanoparticles on human dermal fibroblasts and how to protect cells. *Small* 5 (4):511–520. doi: 10.1002/smll.200800798.

Papir, Y. S., K. H. Hsu, and R. H. Wildnauer. 1975. Mechanical-properties of stratum-corneum. 1. Effect of water and ambient-temperature on tensile properties of newborn rat stratum-corneum. *Biochimica et Biophysica Acta* 399 (1):170–180. doi: 10.1016/0304-4165(75)90223-8.

Patzelt, A., H. Richter, F. Knorr, U. Schafer, C. M. Lehr, L. Dahne, W. Sterry, and J. Lademann. 2011. Selective follicular targeting by modification of the particle sizes. *Journal of Controlled Release* 150 (1):45–48. doi: 10.1016/j.jconrel.2010.11.015.

Pfeifer, G. P. and A. Besaratinia. 2012. UV wavelength-dependent DNA damage and human non-melanoma and melanoma skin cancer. *Photochemical & Photobiological Sciences* 11 (1):90–97. doi: 10.1039/c1pp05144j.

Pinto, A. V., E. L. Deodato, J. S. Cardoso, E. F. Oliveira, S. L. Machado, H. K. Toma, A. C. Leitao, and M. de Padula. 2010. Enzymatic recognition of DNA damage induced by UVB-photosensitized titanium dioxide and biological consequences in *Saccharomyces cerevisiae*: Evidence for oxidatively DNA damage generation. *Mutation Research—Fundamental and Molecular Mechanisms of Mutagenesis* 688 (1–2):3–11. doi: 10.1016/j.mrfmmm.2010.02.003.

Potts, R. O. and R. H. Guy. 1992. Predicting skin permeability. *Pharmaceutical Research* 9 (5):663–669. doi: 10.1023/a:1015810312465.

Raghava, S., M. Hammond, and U. B. Kompella. 2004. Periocular routes for retinal drug delivery. *Expert Opinion on Drug Delivery* 1 (1):99–114. doi: 10.1517/17425247.1.1.99.

Roberts, J. E. 2011. Ultraviolet radiation as a risk factor for cataract and macular degeneration. *Eye & Contact Lens—Science and Clinical Practice* 37 (4):246–249. doi: 10.1097/ICL.0b013e31821cbcc9.

Roberts, J. E., A. R. Wielgus, W. K. Boyes, U. Andley, and C. F. Chignell. 2008. Phototoxicity and cytotoxicity of fullerol in human lens epithelial cells. *Toxicology and Applied Pharmacology* 228 (1):49–58. doi: 10.1016/j.taap.2007.12.010.

Ross, M. H. and W. Pawlina. 2011. *Histology: A Text and Atlas*, 6th edn. Baltimore, MD. Lippincott Williams & Wilkins.

Rozanowska, M., T. Sarna, E. J. Land, and T. G. Truscott. 1999. Free radical scavenging properties of melanin interaction of eu- and pheo-melanin models with reducing and oxidising radicals. *Free Radical Biology and Medicine* 26 (5–6):518–525. doi: 10.1016/s0891-5849(98)00234-2.

Ruddy, S. B. and B. W. Hadzija. 1992. Iontophoretic permeability of polyethylene glycols through hairless rat skin: Application of hydrodynamic theory for hindered transport through liquid-filled pores. *Drug Design and Discovery* 8 (3):207–224.

Ryman-Rasmussen, J. P., J. E. Riviere, and N. A. Monteiro-Riviere. 2006. Penetration of intact skin by quantum dots with diverse physicochemical properties. *Toxicological Sciences* 91 (1):159–165. doi: 10.1093/toxsci/kfj122.

Ryman-Rasmussen, J. P., J. E. Riviere, and N. A. Monteiro-Riviere. 2007. Surface coatings determine cytotoxicity and irritation potential of quantum dot nanoparticles in epidermal keratinocytes. *Journal of Investigative Dermatology* 127 (1):143–153. doi: 10.1038/sj.jid.5700508.

Sadrieh, N., A. M. Wokovich, N. V. Gopee, J. W. Zheng, D. Haines, D. Parmiter, P. H. Siitonen et al. 2010. Lack of significant dermal penetration of titanium dioxide from sunscreen formulations containing nano- and submicron-size TiO_2 particles. *Toxicological Sciences* 115 (1):156–166. doi: 10.1093/toxsci/kfq041.

Sahoo, S. K., F. Diinawaz, and S. Krishnakumar. 2008. Nanotechnology in ocular drug delivery. *Drug Discovery Today* 13 (3–4):144–151. doi: 10.1016/j.drudis.2007.10.021.

Schatzlein, A. and G. Cevc. 1998. Non-uniform cellular packing of the stratum corneum and permeability barrier function of intact skin: A high-resolution confocal laser scanning microscopy study using highly deformable vesicles (Transfersomes). *British Journal of Dermatology* 138 (4):583–592.

Sehedic, D., A. Hardy-Boismartel, C. Couteau, and L. J. M. Coiffard. 2009. Are cosmetic products which include an SPF appropriate for daily use? *Archives of Dermatological Research* 301 (8):603–608. doi: 10.1007/s00403-009-0974-2.

Squier, C. A. and M. J. Kremer. 2001. Biology of oral mucosa and esophagus. *Journal of the National Cancer Institute Monographs* 29:7–15.

Standring, S. 2008. *Gray's Anatomy*, 40th edn. London: Churchill Livingstone, Elsevier.

Timares, L., S. K. Katiyar, and C. A. Elmets. 2008. DNA damage, apoptosis and langerhans cells—Activators of UV-induced immune tolerance. *Photochemistry and Photobiology* 84 (2):422–436. doi: 10.1111/j.1751-1097.2007.00284.x.

Toll, R., U. Jacobi, H. Richter, J. Lademann, H. Schaefer, and U. Blume-Peytavi. 2004. Penetration profile of microspheres in follicular targeting of terminal hair follicles. *Journal of Investigative Dermatology* 123 (1):168–176. doi: 10.1111/j.0022-202X.2004.22717.x.

Vogt, A., B. Combadiere, S. Hadam, K. M. Stieler, J. Lademann, H. Schaefer, B. Autran, W. Sterry, and U. Blume-Peytavi. 2006. 40 nm, but not 750 or 1,500 nm, nanoparticles enter epidermal CD1a+ cells after transcutaneous application on human skin. *Journal of Investigative Dermatology* 126 (6):1316–1322. doi: 10.1038/sj.jid.5700226.

Wakefield, G., S. Lipscomb, E. Holland, and J. Knowland. 2004. The effects of manganese doping on UVA absorption and free radical generation of micronised titanium dioxide and its consequences for the photostability of UVA absorbing organic sunscreen components. *Photochemical & Photobiological Sciences* 3 (7):648–652. doi: 10.1039/b403697b.

Wamer, W. G., J. J. Yin, and R. R. Wei. 1997. Oxidative damage to nucleic acids photosensitized by titanium dioxide. *Free Radical Biology and Medicine* 23 (6):851–858. doi: 10.1016/s0891-5849(97)00068-3.

Wang, S. Q. and I. R. Tooley. 2011. Photoprotection in the era of nanotechnology. *Seminars in Cutaneous Medicine and Surgery* 30 (4):210–213. doi: 10.1016/j.sder.2011.07.006.

Wang, T. F., G. B. Kasting, and J. M. Nitsche. 2006. A multiphase microscopic diffusion model for stratum corneum permeability. I. Formulation, solution, and illustrative results for representative compounds. *Journal of Pharmaceutical Sciences* 95 (3):620–648. doi: 10.1002/jps.20509.

Wang, T. F., G. B. Kasting, and J. M. Nitsche. 2007. A multiphase microscopic diffusion model for stratum corneum permeability. II. Estimation of physicochemical parameters, and application to a large permeability database. *Journal of Pharmaceutical Sciences* 96 (11):3024–3051. doi: 10.1002/jps.20883.

Warner, R. R., M. C. Myers, and D. A. Taylor. 1988. Electron-probe analysis of human-skin-determination of the water concentration profile. *Journal of Investigative Dermatology* 90 (2):218–224. doi: 10.1111/1523-1747.ep12462252.

Wielgus, A. R., B. Zhao, C. F. Chignell, D. N. Hu, and J. E. Roberts. 2010. Phototoxicity and cytotoxicity of fullerol in human retinal pigment epithelial cells. *Toxicology and Applied Pharmacology* 242 (1):79–90. doi: 10.1016/j.taap.2009.09.021.

Wosicka, H. and K. Cal. 2010. Targeting to the hair follicles: Current status and potential. *Journal of Dermatological Science* 57 (2):83–89. doi: 10.1016/j.jdermsci.2009.12.005.

Xia, X. R., N. A. Monteiro-Riviere, and J. E. Riviere. 2010. Skin penetration and kinetics of pristine fullerenes (C-60) topically exposed in industrial organic solvents. *Toxicology and Applied Pharmacology* 242 (1):29–37. doi: 10.1016/j.taap.2009.09.011.

Yamada, Y. 1994. *Modern Textbook of Histology*, 3rd edn. Tokyo, Japan: Kanehara & Co., Ltd.

Yamaguchi, Y., N. Nakamura, T. Nagasawa, A. Kitagawa, K. Matsumoto, Y. Soma, T. Matsuda, M. Mizoguchi, and R. Igarashi. 2006. Enhanced skin regeneration by nanoegg (TM) formulation of all-trans retinoic acid. *Pharmazie* 61 (2):117–121.

Yohn, J. J., M. B. Lyons, and D. A. Norris. 1992. Cultured human melanocytes from black-and-white donors have different sunlight and ultraviolet-A radiation sensitivities. *Journal of Investigative Dermatology* 99 (4):454–459.

Zanolli, M. 2003. The modern paradigm of phototherapy. *Clinics in Dermatology* 21 (5):398–406. doi: 10.1016/j.clindermatol.2003.08.005.

6

Systemic Route

The systemic route includes (1) direct delivery into systemic circulation via the intravenous or buccal route and (2) transfer into systemic circulation after local tissue delivery to the respiratory, GI, integumentary, or sensory systems. Substances injected via the intravenous route circulate through the body and distribute into local tissues via blood vessels. A portion of the locally distributed substance moves to the gastrointestinal (GI) tract via bile excretion or is eliminated outside the body via sweat glands. Intravascularly distributed substances are also rapidly eliminated outside the body from the kidney.

In the pathway of local tissue delivery to systemic distribution, nanoproducts must first pass through the intercellular space and the epithelium of the respiratory, GI, or integumentary system, followed by the endothelial wall of the blood capillary via the basal lamina. In the reverse pathway of systemic circulation to local tissue delivery, nanoproducts first pass through the blood capillary wall and then distribute to the intercellular space and local cells via the basal lamina.

This chapter addresses the anatomy and physiology of systemic tissues and the intercellular space; the transition/translocation of cells; the endocytosis and nuclear transport of nanoproducts; the engineering and practical application of nanoproducts, including the use of nanomedicines to target disease; and the nanoscale world as described by the systemic route.

6.1 Anatomy and Physiology

6.1.1 Intercellular Space

Intercellular space is composed of (1) tight junction complexes that control the permeability of cells and (2) anchoring junctions that link cells and their cytoskeletons to their neighbors, mainly via cadherins and other cell adhesion molecules (Figure 6.1a) (Takeichi, 1991, 1995). Tight junction complexes are critical for epithelial and endothelial permeability via the paracellular pathway of drug delivery. The tight junction complex is composed

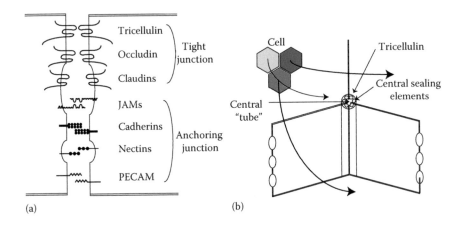

(a)   (b)

FIGURE 6.1
Schematic representations of tight junctions, anchoring junctions, and tricellular junctions.
(a) Tight junctions and anchoring junctions. (Modified from Schneeberger, E.E. and Lynch,
R.D., *Am. J. Physiol.—Cell Physiol.*, 286(6), C1213, 2004, doi: 10.1152/ajpcell.00558.2003; Ebnet, K.,
Histochem. Cell Biol., 130(1), 1, 2008, doi: 10.1007/s00418-008-0418-7; Deli, M.A., *Biochim. Biophys.
Acta*, 1788(4), 892, 2009, doi: 10.1016/j.bbamem.2008.09.016; Hirano, S. and Takeichi, M., *Physiol.
Rev.*, 92(2), 597, 2012, doi: 10.1152/physrev.00014.2011.) (b) Tricellulin in tricellular junctions.
(Redrawn from Ikenouchi, J. et al., *J. Cell Biol.*, 171(6), 939, 2005, doi: 10.1083/jcb.200510043.)

of parameters A–B. These parameters all crucially affect paracellular permeability when overexpressed/stimulated or downregulated/inhibited.

A. Proteins with four membrane-spanning domains (claudin, occludin,
   and tricellulin)

B. Proteins with one membrane-spanning domain (junctional adhesion molecules)

C. The actin cytoskeleton and scaffolding proteins linking the cell
   membrane to the cytoskeleton, including coxsackie/adenovirus-
   associated receptor and zona occludens 1, 2, and 3

D. Transcription factors, kinases, phosphatases, and other signaling
   proteins

Professor Shoichiro Tsukita initially discovered tight junction complexes,
and Furuse et al. went on to identify occludin and claudin (Furuse et al.,
1993, 1998a,b). Itoh et al. showed that occludin and claudin are directly linked
in tight junctions with zona occludens 1, 2, and 3 (Itoh et al., 1993, 1999),
and Ikenouchi et al. (2005) showed that tricellulin forms a novel barrier at
tricellular contacts between three epithelial cells (Figure 6.1b) (Ikenouchi
et al., 2005).

Claudin has 24 isoforms, corresponding to claudins 1–9, 10A, 10B, 11–17, 18-1,
18-2, and 19–24 (Tsukita, 2001; Tsukita et al., 2001; Van Italie and Anderson,

2006; Yamazaki et al., 2011). Claudins allow cell-type-specific paracellular permselectivity, whereas other tight junction proteins are shared by multiple cell types (Angelow et al., 2008). For example, different claudin isomers are expressed by each tissue in the kidney and the lung (Angelow et al., 2008; Soini, 2011). The size of the paracellular pore in claudin-based tight junctions has not yet been determined for all isoforms. Van Itallie et al. (2008) conducted a permselectivity study of paracellular pores in Caco-2 human epithelial colorectal adenocarcinoma cells and pig ileum by using polyethylene glycol (PEG) particles. The results of this study revealed a cutoff substance transport size of ~1.4 nm, while overexpression of claudin 2 modified the pore size to ~0.7 nm.

6.1.2 Basal Lamina

Local tissue delivery to systemic circulation necessitates that substances cross into blood capillaries via the basal lamina/basement membrane. The basal lamina, or lamina densa, forms a 40–60 nm thick layer between the epithelium and neighboring connective tissue and is composed of laminin, type IV collagen, and various integrated proteoglycans and glycoproteins (Figure 6.2a) (Ross and Pawlina, 2011). The thickness of the basal lamina changes with tissue location and in some cases includes a layer of fibroreticular tissue (Figure 6.2b). The fibroreticular tissue contains a network of self-assembled type IV collagen fibrils. Therefore, passage to the blood capillary via the epithelium requires that a substance transverse the laminin/collagen suprastructure.

Laminin is an α-helical coiled coil, heterotrimeric glycoprotein with a cross-like structure that is composed of an α, a β, and a γ chain (Figure 6.2c) (Timpl and Brown, 1994; Colognate et al., 1999; Ekblom et al., 2003; Hallmann et al., 2005; Scheele et al., 2007; Tzu and Marinkovich, 2008; Durbeej, 2010). Like collagen IV, laminin shows a reticulated pattern due to self-assembly via its terminal amino groups (Figure 6.2b), while the terminal carboxyl groups interact with and connect the cell membrane to the basement membrane.

Forty-five laminin isoforms can potentially be generated by various combinations of five types of α, three types of β, and three types of γ chains. Of these, 18 naturally occurring isoforms have been identified that show tissue-selective expression patterns. Laminin-111 is expressed in blood vessel walls in the brain and shows an epithelial localization during fetal embryogenesis and in the mature kidney, liver, testis, and ovary. Laminin-511 is expressed in the epithelium during embryogenesis and in the mature epithelium, endothelium, and smooth muscle. Laminin-323 is expressed in the central nervous system and in reticular cells. In addition to their structural roles, assorted laminin isoforms participate in intracellular signaling pathways stimulated by the extracellular matrix/basement membrane.

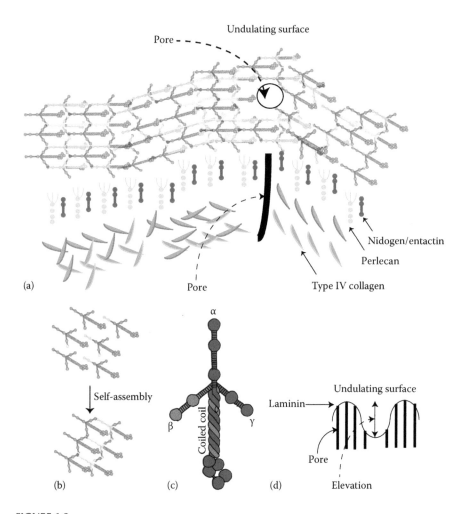

FIGURE 6.2
Schematic representation of the basement membrane. (a) Basement membrane, (b) self-assembly of basal lamina with reticular layer, (c) structure of laminin, and (d) undulating surface and pore of the basement membrane. (Modified from Ross, M.H. and Pawlina, W., *Histology: A Text and Atlas*, 5th edn., Lippincott Williams & Wilkins, Baltimore, MD, 2006; Abrams, G.A. et al., *Cornea*, 19(1), 57, 2000b; Durbeej, M., *Cell Tissue Res.*, 339(1), 259, 2010, doi: 10.1007/s00441-009-0838-2.)

Passage of macromolecules through the basement membrane is influenced by fiber pore size and permeability via laminin. This size is 200 nm or less in the corneal basement membrane. For instance, scanning electron microscopy (SEM) analysis revealed a mean pore diameter and an interpore distance of 72 and 87 nm, respectively, in the anterior corneal basement membrane of the macaque monkey, while the pore diameter ranged from 22 to 216 nm (Abrams et al., 2000a). The mean pore diameter was 82 nm and the interpore distance was 127 nm in the macaque bladder epithelial basement membrane, while the pore diameter ranged from 13 to 222 nm (Abrams et al., 2003).

The basement membrane surface is undulating rather than planar (Figure 6.2d). Fiber and pore diameters and elevation features were investigated in an additional SEM analysis of the anterior corneal basement membrane and the corneal Descemet's membrane in humans. The mean pore diameter and elevation in the anterior corneal basement membrane were 92 nm (range, 30–191 nm) and 182 nm (range, 111–267 nm), respectively; the corresponding values were 38 nm (range, 22–87 nm) and 131 nm (range, 79–229 nm) in Descemet's membrane (Abrams et al., 2000b). Meanwhile, pore sizes in the aorta, carotid, saphenous, and inferior vena cava in the rhesus macaque were 59, 63, 38, and 49 nm, respectively (Liliensiek et al., 2009). SEM also revealed that the pores in the glomerular basement membrane ranged from 11 to 30 nm in diameter (Shirato et al., 1991).

Biophysical cues provided by the substrate modulus in the basal lamina differ by each tissue. In the human cornea, the elastic modulus of the anterior basement membrane (7.5 kPa) and Descemet's membrane (50 kPa), as measured by AFM (Last et al., 2009), revealed a more tightly packed structure for Descemet's membrane compared with the anterior basement membrane.

6.1.3 Blood Vessels

Blood flows from the heart to the pulmonary artery and returns to the heart via the pulmonary vein. Substances in the blood must travel through the intracellular pathway of endothelial pores or the paracellular pathway of endothelial gaps in the basement membrane/extracellular matrix to move across the capillaries and into local tissue. Substances in local tissue must similarly pass through endothelial pores or gaps in the basement membrane/extracellular matrix to enter the vasculature. The anatomical structure of capillaries in various parts of the body therefore strongly affects endothelial permeability (Figure 6.3).

Capillaries in the central nervous system or muscle are formed by a closed or continuous endothelium. The pores or gaps in the continuous endothelium are smaller in size and fewer in number than in other endothelia. Capillary properties are especially remarkable in the brain and contribute to the blood–brain barrier. The average number of plasma membrane vesicles (pores) in the heart muscle is 89/mm^2, while that in the brain is only 5/mm^2 (Tani et al., 1977).

Capillaries in the glomerular endothelium of the kidney and the main endocrine system are of the fenestrated type. The fenestra, or pore, develops from the sieve plate. Pores in the renal glomerular endothelium are round in shape and 60 nm in diameter and are therefore larger than those in the closed or continuous endothelium. The pores are compartmentalized by a

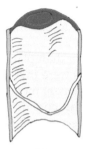

Closed or continuous type

Organ: central nervous system and muscle
Size of gap: <1 nm

Sieve plate

Kidney type or pore type

Organ: the part of endocrine system and kidney
Size of pore: 60 nm
The fenestrae are compartmentalized by diaphragm
Shape of permeable pore: fan shape
Size of permeable pore: around 5 nm

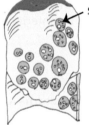

Sieve plate

Liver type

Organ: the part of endocrine system as liver or
       adrenal or bone marrow
Fenestrae or pore is lacking diaphragm
Size of pore: <100 nm (70%), 10–500 nm (28%),
       >500 nm (2%)
Size of gap: 1000 nm–3000 nm

Spleen type

Organ: spleen, lymph
For an example of spleen, capillary is formed by
       the rod
Size of gap: 1000 nm–2000 nm

FIGURE 6.3
Schematic representation of blood capillaries. (Modified from Fujita, H. and Fujita, T., *Textbook of Histology: Part 2*, 3rd edn., Igaku-Shoin Ltd., Tokyo, Japan, 1992; Yamada (1994).)

fibrin-based diaphragm, and the shape of the compartment or communication channel resembles that of a fan.

Capillaries in the liver, adrenal endocrine system, and bone marrow have large pores and gaps. The pores develop from the sieve plate, are round in shape, and lack the fibrin diaphragm. The hepatic endothelium contains pores or gaps of four sizes, ≤100 nm (70% of all pores), 100–500 nm, and ≥500 nm, including gaps of 1–3 µm (Ishimura et al., 1978; Fujita and Fujita, 1992; Igarashi, 2008a).

Lastly, capillaries in the spleen are formed by rod cells and have large gaps of ~1–2 µm in diameter.

6.2 Pathway from Systemic Exposure to Local Delivery

6.2.1 Renal Blood Capillary Permeability

Renal glomerular filtration through the Bowman capsule to the ureter is the first step in removing low-molecular-weight (LMW) substances from the blood into the urine. Glomerular filtration involves three separate barriers: the fenestrated capillary endothelium, the glomerular basement membrane, and the podocyte (visceral epithelial cell) layer (Figure 6.4a through d). The capillary endothelium is the major contributor to the glomerular filtration barrier (Ballermann and Stan, 2007). The glomerular endothelial pore is 60 nm on average and is compartmentalized by a fibrin diaphragm (Maul, 1971). Although the diaphragm cannot be detected by conventional methods, it is revealed by improved methods of specimen processing, such as freeze fracture (Figure 6.4e) (Bearer et al., 1985). The pore is composed of fibrins (16 fibrin molecules [2 sets of 8] in theory, 15 on average) (Bearer et al., 1985), and glycan antennae are located around the end of the fibrin diaphragm (Figure 6.5) (Stan, 2007).

The penetration of substances through the renal endothelial barrier depends on their size, shape, and charge (Bearer et al., 1985). Other factors include the radius of the fan-shaped pore, fibrin width, and number of fibrin molecules (Igarashi, 2008b) (Table 6.1). Igarashi (2008b) estimated the size of spherical, uncharged nanoparticles that can pass through the pore as ~5 nm, based on a consideration of pore constituents as either the theoretical set of 16 fibrin molecules or the average set of 15 molecules reported by Bearer et al. (1985). Because the pore is fan shaped rather than circular, spherical nanoparticles that pass through the channel will be smaller than the actual channel circumference (4.8 and 5.4 nm for a 16-fibrin channel and a 15-fibrin channel, respectively). Unfortunately, the space taken up by the glycan antennae was not included in this calculation. Nonetheless, the estimated particle size of 5 nm is in good agreement with a particle size of ≤5.5 nm revealed by quantum dot analysis (Choi et al., 2007; Landsiedel et al., 2012).

Podocytes wrap around the capillaries of the glomerulus, leaving slits between the cells (or slit diaphragms) adjacent to the capillary basement membrane (Fujita et al., 1970; Rodewald and Karnovsky, 1974). The slit diaphragm is formed by associations between multiple cell surface proteins and consists of a

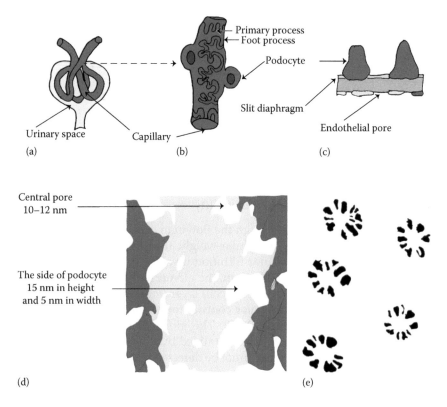

FIGURE 6.4
Permeability barriers in glomerular filtration. (a) Glomerulus. (Modified from Standring (2008).)
(b) Glomerular capillary. (Modified from Standring, S., *Gray's Anatomy*, 40th edn., Churchill
Livingstone, Elsevier, London, 2008.) (c) Filtration barrier of slit diaphragm. (Modified from
Ross, M.H. and Pawlina, W., *Histology: A Text and Atlas*, 5th edn., Lippincott Williams &
Wilkins, Baltimore, MD, 2006.) (d) Slit diaphragm pore. (Modified from Supplement 1B of
Wartiovaara, J. et al., *J. Clin. Invest.*, 114(10), 1475, 2004, doi: 10.1172/jci200422562.) (e) Endothelial
pore. (Modified for only pore from Bearer, E.L. et al., *J. Cell Biol.*, 100(2), 418, 1985, doi: 10.1083/
jcb.100.2.418.)

longitudinal zipper-like chain and a central pore. The central pore is 10–12 nm
wide and narrows to a width of 5 nm and a height of 15 nm on the side facing
the podocyte (Figure 6.4d) (Wartiovaara et al., 2004). SEM analysis showed
that the average pore diameter in the rat glomerular basement membrane
was 14.1 nm and ranged from 4 to 120 nm (Hironaka et al., 1993). Given the
relatively large size of the pore in the slit diaphragm, this structure may not
restrict the passage of nanoparticles through the glomerular filtration barrier.

Drug disposition in the kidney and renal elimination were analyzed by
using 1–15 nm gadolinium chelate-labeled, poly(amidoamine) dendrimer
contrast agents (Kobayashi and Brechbiel, 2005). Dendrimers are polymeric
branching molecules with a treelike structure. A polymer arm that branches

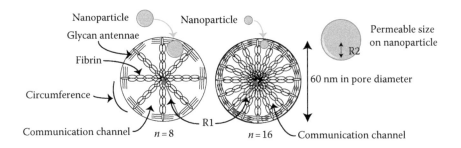

FIGURE 6.5
Renal blood capillary showing endothelial pore size and influx of a nanoparticle into the communicating channel. (Adapted from Igarashi, E., Factors affecting toxicity and efficacy of injectable nanomedicines: Significance on particle size, in: *Paper Read at EHRLICH II—Second World Conference on Magic Bullets, Celebrating the 100th Anniversary of the Nobel Prize Award to Paul Ehrlich*, May 15, Nürnberg, Germany, 2008b.)

TABLE 6.1

Endothelial Pore Size Based on Assumed Number of Fibrin Molecules

Assumptions by Fibrin Number (n: Octagon)	Radius (R1) of Fenestral Pores (nm)	Fibrin Width (nm)	Fibrin Number	Circumference (nm)	Radius (R2) of Circle in Communicating Channel (nm)
Observed number in few case	30	6	9	14.93	5.45
Observed number in few case	30	6	12	9.70	3.81
Average of observed number	30	6	15	6.56	2.70
Theoretical number (n=2)	30	6	16	5.78	2.40

Source: Adapted from Igarashi, E., Factors affecting toxicity and efficacy of injectable nanomedicines: Significance on particle size, in: *Paper Read at EHRLICH II—Second World Conference on Magic Bullets, Celebrating the 100th Anniversary of the Nobel Prize Award to Paul Ehrlich*, May 15, Nürnberg, Germany, 2008b.

Note: R1 is the radius of fenestral pores, while R2 is the radius of permeable size on nanoparticle.

into three directions is referred to as generation one (G1). Dendrimers are classified by the number of generations as G1, G2, G3, G4, G5, and so on. Dendrimer technology allows precise size control in the nanoscale and produces nanoparticles that differ by 1–2 nm for each generation. Kobayashi and Brechbiel (2005) showed that the renal excretion rate decreased as dendrimer size increased. Dendrimers of ≤3 or 3–6 nm passed through the capillaries after intravenous injection and were rapidly eliminated by the kidney. Dendrimers of 7–12 nm were mainly eliminated by the liver, showing minimal renal excretion and high uptake by the reticuloendothelial system.

The manner in which dendrimers are manufactured decrees that the number of infinite and irregular forms should increase with increasing numbers of generations. Naturally occurring dendrimers exhibit elliptical rather than spherical shapes. Because nanoparticle permeance is determined by the shortest diameter that can pass through a pore, the diameter of an infinite form or ellipse inaccurately reflects the cutoff size. On the other hand, the renal glomerulus does not eliminate plasma albumin in the blood, which has an ellipsoid-disk shape and is 8.2 × 6.3 × 2.75 nm sized (three dimensionally) (Anderegg et al., 1955), or ~7.2 nm in diameter by the Stokes–Einstein radius (Haraldsson et al., 2008). The estimated cutoff size for renal elimination of a dendrimer is therefore assumed to be smaller than the shortest diameter of albumin, or 6.3 nm.

Ferritin is a 12–13 nm doughnut-shaped, iron storage protein that can store 4500 iron atoms in a central 7–9 nm cavity (Harrison and Arosio, 1996; Aisen et al., 2001; Liu et al., 2003). The binding activity of ferritin was studied at the surface of the glomerular epithelium by using a modified form of ferritin imbued with electrostatic properties. Studies of charge-decorated nanoparticles demonstrated a fusion event between cationic ferritin and plasmalemmal vesicle-associated protein-1 (PV-1) (Figure 6.6), while anionic or neutral ferritin did not bind with PV-1 (Koshy and Avasthi, 1987; Avasthi and Koshy, 1988a,b; Ballermann and Stan, 2007). PV-1 is an important constituent of

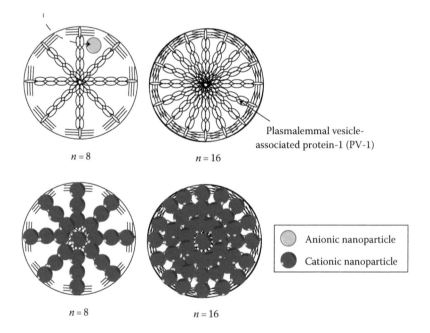

$n = 8$ $n = 16$

Plasmalemmal vesicle-associated protein-1 (PV-1)

$n = 8$ $n = 16$

○ Anionic nanoparticle
● Cationic nanoparticle

FIGURE 6.6
Mechanism of occlusion by nanoparticle binding with PV-1.

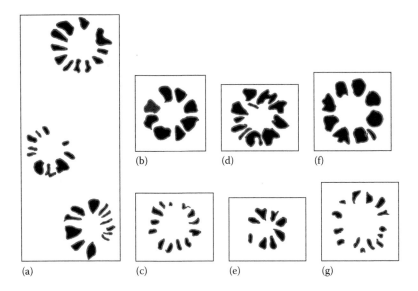

FIGURE 6.7
Endothelial pores compartmentalized by fibrin in various kinds of renal tissue. (a) Glomerulus, (b, c) adrenal cortex, (d) endocrine pancreas, (e) exocrine pancreas, and (f, g) renal peritubular capillaries. (Pore part was modified from Bearer, E.L. et al., *J. Cell Biol.*, 100(2), 418, 1985, doi: 10.1083/jcb.100.2.418.)

the fibrin-based diaphragm in renal glomerular capillary endothelial pores (Stan, 2004; Stan et al., 2004) and is fundamental to membrane permeability and nanoparticle disposition. Injection of copious amounts of cationic amino acids is acutely lethal to mammalian animals (Dickerman and Walker, 1964), and the lethal effect is attributed to occlusion of capillary pores by amino acid–induced clot formation and excessive PV-1 binding with cationic ferritin (Figure 6.6) (Avasthi and Koshy, 1988b).

Endothelial pores symmetrically compartmentalized by fibrin also exist in capillaries of the endocrine system (adrenal cortex, endocrine pancreas, and exocrine pancreas) and in the renal peritubular capillaries (Figure 6.7) (Bearer et al., 1985).

6.2.2 Permeability of the Liver and Splenic Sinus

Endothelial pores in sinusoidal hepatic capillaries are spherical in shape and lack the fibrin diaphragm (Ballermann, 2005). This type of endothelium is termed the "liver type" and is also found in bone marrow and the adrenal cortex (Muto, 1976; Ishimura et al., 1978). In mice, 70% of the hepatic endothelial pores are small in size (≤100 nm), 28% are intermediate (100–500 nm), and 3% are large (≥500 nm) (Ishimura et al., 1978). Humans and mice exhibit a similar distribution of small, intermediate, and large pores (Ishimura et al., 1978; Horn et al., 1986).

Spleen- and lymph-type capillary endothelia share a common mechanism of opening and closing endothelial microvalves, with periodic expansion and compression (Schmid-Schonbein, 1990; Ross and Pawlina, 2011). The spleen-type endothelial gap is ~1–2 μm in diameter. Therefore, an intravenously injected, 5–100 nm nanoparticle should be circulated through the sinusoidal space (the space of Disse) via liver-type capillaries and then through the splenic sinus via gaps formed by rod cell constituents of spleen-type capillaries during periodic expansion. The same nanoparticle should not be distributed to the heart, lung, kidney, central nervous system, or muscle. This form of nanoparticle distribution was demonstrated in a pharmacokinetic study by using 30 nm neutral or anionic polymeric nanomedicines (Uchino et al., 2005). However, intravenously injected 100–200 nm nanoparticles should be circulated in the vasculature through the sinusoidal space in the liver, as well as through the lymphatic and splenic sinuses. This hypothesis was supported by additional pharmacokinetic studies using 150 nm anionic liposomes (Phan et al., 2005; Zolnik et al., 2008).

Liposomes include soft and hard types, depending on lipid constitution. Liposome rigidity affects the capacity to penetrate into membranes and subsequent distribution. For example, multilamellar 200 nm liposomes were coated with soybean-derived sitosterol glucoside (SG) as a ligand targeting liver parenchymal cells and injected into mice. The liposomes interacted with hepatic parenchymal cells, and 50% of the injected volume accumulated in the liver within 1 h of administration. However, the liposomes did not accumulate in the spleen (Maitani, 1996; Shimizu et al., 1996). The distribution in parenchymal liver cells was about sevenfold higher than in nonparenchymal liver cells, such as Kupffer cells (Figure 6.8). Because 200 nm nanoparticles do not easily pass through 70% of the hepatic endothelial pores, the liposomes should hypothetically have accumulated in capillaries and splenic and lymphatic tissues, rather than in the liver.

The disparity between the anticipated and the actual results are explained by liposome rigidity. The SG-coated liposomes described earlier are categorized as soft and can change their shape such that the shortest diameter is ≤100 nm rather than 200 nm, and thus the aspect ratio is drastically altered (Figure 6.9). On the other hand, doxorubicin-loaded liposomes of ~100 nm (Doxil®) are stable in the blood (half-life, 55 h) and accumulate in the spleen rather than the liver after intravenous injection in humans. Therefore, Doxil liposomes are categorized as hard. The disposition of hard versus soft liposomes suggests that the particle size measured at the time of manufacture is not necessarily a good predictor of particle disposition; the shortest length and aspect ratio are more critical parameters.

The liver, bone marrow, and certain endocrine tissues have large endothelial pores. These tissues play an active role in the production of differentiated blood cells or physiologically active substances (e.g., endocrine hormones) that circulate in the blood. Blood flow through these tissues occurs in two directions, as inflow and outflow, and nanoparticles tend to accumulate in

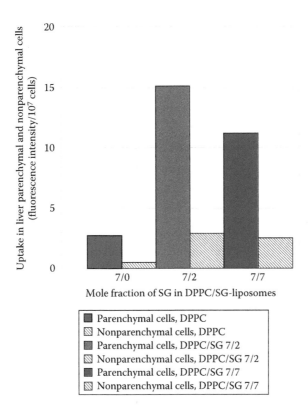

FIGURE 6.8
Distribution of liposomes in hepatic cells at 2 h after intravenous injection in mice. Dark-colored bars, parenchymal cells; light-colored bars, nonparenchymal cells (red, dipalmitoylphosphatidylcholine, DPPC, without SG; green, 7:2 ratio of SG to DPPC; blue, 7:7 ratio of SG to DPPC). Each value represents the mean (n = 3). (Redrawn from data in Maitani, Y., *Yakugaku Zasshi—J. Pharmaceut. Soc. Jpn.*, 116(12), 901, 1996.)

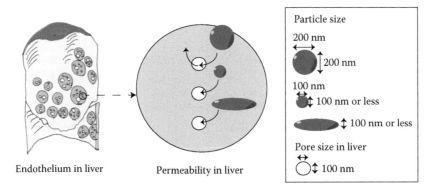

FIGURE 6.9
Relationship between particle and pore size in the liver.

such tissues. However, distribution results are affected by the volume of the blood vessels and environmental cues. For example, liver endothelial cells respond to environmental changes in vitro by controlling inflow and outflow via the opening or closing of endothelial pores (Braet et al., 1995). The average pore size increases by 5% in response to exogenous alcohol and decreases by 20% in response to exogenous serotonin.

6.2.3 Nanoparticle Interactions with Proteins and Cells

The interaction of a nanoparticle with plasma proteins and other blood constituents affects its disposition and hematocompatibility (i.e., thrombogenicity, as well as the capacity to provoke hemolysis, complement activation, and opsonophagocytic activity). Dobrovoskaia et al. (2008) reviewed the relationship between nanoparticle properties and hematocompatibility and reported that strongly cationic fullerenes and dendrimers induced hemolysis via direct interaction with erythrocyte membranes, whereas anionic nanoparticles did not.

Conventional cationic substances are less likely to provoke such a response because they are rapidly eliminated by renal excretion. By contrast, an intact nanoparticle escaping renal excretion displays a longer time in blood circulation, increasing the prospect of interactions with blood components and naturally occurring anticoagulants. Interactions with anticoagulants are particularly harmful due to partial or complete obstruction of the blood vessel following nanoparticle-mediated activation of the anticoagulation cascade, blood coagulation, and thrombus/blood clot formation.

The mechanism of blood clot formation induced by nanoparticles differs from that induced by microparticles. For charged nanoparticles that directly damage the erythrocyte membrane, blood clot formation depends on the strength of the surface charge. For instance, quenching of a positively charged nanoparticle by the addition of a negatively charged PEG coating decreases its capacity to trigger platelet aggregation and blood clotting. For nanoparticles that do not directly damage the membrane (e.g., water-soluble, charged fullerenes; single-walled carbon nanotubes; and multiwalled carbon nanotubes [MWCNTs]), platelet aggregation and blood clot formation are initiated through the activation of a receptor for glycoprotein IIb/IIIa. By contrast, protein kinase activation is required for microparticle- but not nanoparticle-stimulated platelet aggregation.

A complement is a group of proteins linked together in a biochemical cascade that removes pathogens from the body by cellular and humor immunity. Hypersensitivity reactions and anaphylactic shock are attributed to complement activation after exposure to immunogenic substances. Cationic nanoparticles can induce complement activation, but this capacity is lost by PEG or poloxamine-908 coating. PEG-coated doxorubicin liposomes (Doxil) engendered hypersensitivity and complement activation in clinical studies although PEG is itself negatively charged (Chanan-Khan et al., 2003).

Cationic and strongly anionic nanoparticles also stimulate opsono-phagocytosis more readily than neutral or weakly anionic nanoparticles. Nanoparticle binding to plasma proteins initiates the formation of a plasma protein–coated nanoparticle complex termed a protein corona that is much larger than the intact nanoparticle (Aggarwal et al., 2009; Casals et al., 2010; Montes-Burgos et al., 2010; Tenzer et al., 2011; Zhang et al., 2011a). Protein coronas are easily recognized and endocytosed by macrophages or Kupffer cells (Aggarwal et al., 2009).

The effect of nanoparticle size and charge modification in blood was examined on complement activation for two gold colloidal nanoparticles: a 32 nm nanoparticle with an anionic charge of −38.2 mV and a 55 nm nanoparticle with an anionic charge of −33.4 mV (Dobrovolskaia et al., 2009). The nanoparticle surface charge was weakened to −16.4 mV for the 32 nm particle and −17.6 mV for the 55 nm particle, which in turn modified the average sizes of the nanoparticles from 32 to 76.7 nm and from 55 to 100.0 nm, respectively, as determined by DLS. The weakly anionic gold nanoparticles failed to bind to plasma albumin or trigger complement activation, although they did reversibly bind to proteins closely associated with blood coagulation and platelet aggregation (e.g., fibrinogen).

Fibrinogen does not aggregate under normal physiological conditions. The protein only aggregates due to delayed circulation caused by external injury or attenuated cardiac function. Fibrinogen aggregation then contributes to wound protection and healing by incrustation or to cerebral infarction by blood clotting. Therefore, two probabilities arise for nanoparticle fate in the blood. Strongly anionic nanoparticles are predicted to tightly bind to fibrinogen and trigger blood clotting in slow-moving blood, whereas weakly anionic nanoparticles are expected to reach equilibrium at −16 or −17 mV.

Recently, a biological surface adsorption index was proposed to reflect the adsorption strength of nanoparticles with proteins for the generation of protein coronas (Xia et al., 2010). This index was utilized to map the adsorption strength of 15 nanoparticles (Xia et al., 2011). Static metallic nanoparticles composed of silica dioxide or titanium dioxide showed weak adsorption, while fullerenes showed intermediate adsorption, and MCNTs showed the strongest adsorption of all. These unexpected results suggest that the adsorption power of numerous nonspecific van der Waals forces is stronger than that of electrostatic forces.

Adsorption and desorption are both expected to take place during the formation of protein coronas. Adsorption and desorption are extremely rapid processes that occur in milliseconds, according to simulation studies of surface-acting agents (Frise et al., 2010; Obata and Honda, 2011). Hence, MCNTs are expected to show similar behavior in protein corona formation. Although strongly charged nanoparticles also induce the production of protein coronas, Wang et al. (2011) showed that weaker van der Waals forces were more likely to alter the function of enzymes and proteins in the protein complex (Wang et al., 2011).

When stealthy nanoparticles (i.e., those that escape detection by the immune system) move at high speeds through the blood stream, their ability to travel via transcellular pathways depends on the shortest diameter of the nanoparticle and the size and shape of the endothelial pore. However, the paracellular permeance of a nanoparticle/plasma protein complex depends on its entire surface area.

Most intravenously injected, stealthy nanoparticles are circulated through the capillaries to the liver and the spleen, but they escape distribution to the brain because of the blood–brain barrier. Metallic nanoparticles sometimes distribute to the brain (Sharma et al., 2009, 2010), depending on the metal. Silver and copper nanoparticles show a higher distribution to the brain than aluminum nanoparticles in rats and mice.

6.2.4 Basement Membrane Permeability and Convection Effects

The basement membrane/basal lamina displays a regular netlike structure generated by the self-assembly of laminin and type IV collagen, with randomly dispersed fibrils of laminin, collagen, and proteoglycan. Therefore, specialized pores are not expected, although nanoscale pores can be observed in the basement membrane of the bladder, cornea (Abrams et al., 2000a,b, 2003), renal glomerulus (Shirato et al., 1991), aorta, carotid, saphenous vein, and inferior vena cava (Liliensiek et al, 2009) (also see Section 6.1.2). The presence of these pores suggests that nanoparticles may cross the basement membrane of certain endothelial or epithelial cells via convection-enhanced transport.

To investigate this possibility, the biodistribution of irinotecan as an entrapped, water-soluble marker inside organic nanotubes was examined at 3, 24, and 48 h after intravenous injection in mice. Cylindrical organic nanotubes (inner diameter, 30–40 nm; outer diameter, 70–90 nm; length, ~2 μm) coated with gadolinium or loaded with irinotecan were utilized in the study, as well as 3 μm spherical microparticles (Maitani et al., 2013). Irinotecan-loaded nanotubes showed a higher distribution to the lung and spleen than free irinotecan at 24 h postinjection, with accumulation of the active irinotecan metabolite, SN-38, in the lung and liver. The gadolinium-coated nanotubes largely accumulated in the lung.

The 3 μm microparticles exhibited high accumulation in the liver and spleen relative to both kinds of nanoproducts at 3 and 24 h postinjection but rarely accumulated in the lung. Most of the microparticles were eliminated after 48 h. The results with microparticles and high-aspect-ratio nanotubes suggest that the lung has special permeability characteristics. This idea is supported by SEM and transmission electron microscopy (TEM) observations of oval-shaped pores in the basement membrane of the human bronchial airway obtained from postsurgery bronchial mucosa specimens and the fact that the airway can be used as a conduit for immune cells to traffic between the epithelial and mesenchymal compartments (Howat et al., 2001).

SEM micrographs indicated that the mean pore diameter and pore count per unit area in the bronchial basement membrane were 1.76 μm (range, 0.6–3.85 μm) and 863 pores/mm^2 (range, 208–2337 pores/mm^2). In addition, TEM sections showed that the mean pore diameter and distance between a pore and its nearest neighbor were 2.08 μm (range, 0.62–4.2 μm) and 12.8 μm (range, 4.0–31.8 μm), respectively. The high accumulation of organic nano-tubes in the lung is therefore consistent with the anatomical pore size of the basement membrane and the existence of microscale convection effects in the bronchial basement membrane.

6.3 Endocytosis of Nanoparticles

G protein–coupled receptors (GPCRs) form a family of receptors with high tissue selectivity and cover the largest cell membrane surface area of any receptor family in the human genome (Kobilka, 2007; Lefkowitz, 2007; Lagerstrom and Schioth, 2008). For example, there are only 2 types of transferrin receptor (Qian et al., 2002), while the GPCRs comprise 7 receptor families and 799 genes in the human (Gloriam et al., 2007). Approximately 30% of marketed drugs target GPCRs (Kontoyianni and Liu, 2012).

Coating of nanoparticles with GPCR-selective ligands permits endocytosis via GPCR family members. This phenomenon is exemplified by octreotide, an octapeptide that pharmacologically mimics somatostatin. Octreotide-coated 150 nm liposomes and 80 nm polymeric micelles are both successfully endocytosed by tumor cells expressing the somatostatin GPCR (Iwase and Maitani, 2011; Zhang et al., 2011b). Furthermore, chemically coated 27.5 nm quantum dots are endocytosed via low-density lipoprotein GPCRs (Zhang and Monteiro-Riviere, 2009).

The human genome contains 799 GPCR family genes, compared with 1783 in mice and 1277 in rats (Gloriam et al., 2007). These species-specific differences have been mainly attributed to the high number of olfactory bulb GPCRs in nonhuman animals and suggest that endocytosis via GPCRs occurs more frequently in experimental animals such as mice or rats than in humans.

6.4 Nuclear Transport

Intracellular nuclear transport is mediated via constituents of the nuclear pore complex (NPC), namely, importin for transport from the cytoplasm into the nucleus and exportin for transport from the nucleus into the cytoplasm (Figure 6.10a). The NPC has an outer diameter of 125 nm, a height

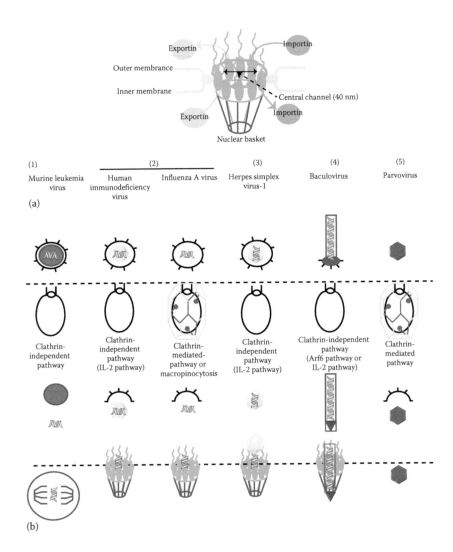

FIGURE 6.10
Mechanisms of endocytosis and intracellular nuclear entry of virus particles. (a) The NPC and (b) mechanisms of virus invasion. (Modified from Cohen, S. et al., *Biochim. Biophys. Acta—Mol. Cell Res.*, 1813(9), 1634, 2011, doi: 10.1016/j.bbamcr.2010.12.009; Friedrich, B.M. et al., *Virus Res.*, 161(2), 101, 2011, doi: 10.1016/j.virusres.2011.08.001; Spear, M. et al., *J. Biol. Chem.*, 289(10), 6949, 2014, doi: 10.1074/jbc.M113.492132; Wayengera, M., *Theor. Biol. Med. Modell.*, 7, 5, 2010, doi: 10.1186/1742-4682-7-5; Edinger, T.O. et al., *J. Gen. Virol.*, 95, 263, 2014, doi: 10.1099/Vir.0.059477-0; Favoreel, H.W. et al., *Trends Microbiol.*, 15(9), 426, 2007, doi: 10.1016/j.tim.2007.08.003; Laakkonen, J.P. et al., *PLoS ONE*, 4(4), E5093, 2009, doi: 10.1371/Journal.Pone.0005093; Mudhakir, D. and Harashima, H., *AAPS J.*, 11(1), 65, 2009, doi: 10.1208/s12248-009-9080-9.)

of 50–70 nm, and a central channel diameter via active translocation by transport receptors of ~39 nm. LMW substances of ≤40 kDa (or 9 nm) cross the NPC via passive diffusion, whereas HMW substances cross via active transport, an energy-dependent process requiring stimulation of signal transduction pathways (Peter, 2009; Yasuhara et al., 2009; Cohen et al., 2011; Hoelz et al., 2011). Gu et al. (2009) examined nuclear transport of 3.7 nm gold nanoparticles in HeLa cells. The nanoparticles were endocytosed at 6 h after administration and reached equilibrium at 24 h, showing partial nuclear localization.

Intracellular nuclear invasion by virus particles provides important information regarding the nuclear transport of nanoparticles, including the size and shape of nanoparticles that enable such transport. The nuclear transport of virus particles takes place by the following five scenarios (Figure 6.10b) (Cohen et al., 2011): (1) The murine leukemia virus accesses the cellular nucleus during mitotic cell division. (2) Human immunodeficiency virus 1 and influenza A virus release proteins required for virus genome (DNA) replication into the cytoplasm, which then target the nucleus via nuclear localization sequences. (3) Herpes simplex virus-1 capsids (180–200 nm in diameter) are released into the cytoplasm, and the virus genome (DNA) crosses the NPC and is released into the nucleus. (4) Hepatitis B virus and the baculoviruses cross the NPC and release the virus genome (DNA) into the nucleus (Passrelli, 2011). (5) Parvoviruses generate temporary damage to the nuclear layer for passage into the nucleus.

6.5 Cell Transition/Translocation

Nanoparticles translocate from local tissue to systemic tissue and/or from systemic tissue to local tissue. Bacteria and viruses also translocate throughout the body, as do metastatic cancer cells and normal cells under conditions of injury and inflammation. For example, necrotic tissue is replaced by normal tissue during wound healing in a process that requires cell migration, while metastatic cancer cells travel through the vasculature from one site to the next. Furthermore, epithelial cells gain migratory and invasive properties during epithelial–mesenchymal transition, whereas mesenchymal cells lose their motility during mesenchymal–epithelial transition (Figure 6.11). Both forms of cell transition occur in normal development as well as metastatic cancer. Epithelial–mesenchymal transition is also a hallmark of idiopathic pulmonary fibrosis.

Bacterial flora in the intestine control translocation via tight junctions and the paracellular pathway by increasing or decreasing the expression of tight junction proteins in the intestinal epithelium (Ohland and MacNaughton, 2010; Ulluwishewa et al., 2011). *Escherichia coli, Bifidobacterium, Helicobacter*

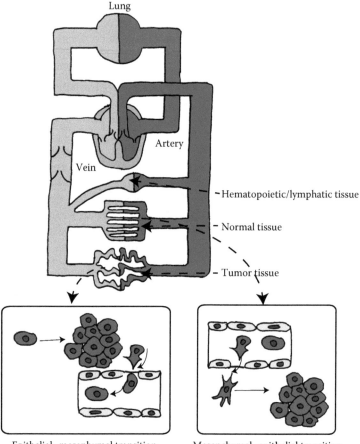

FIGURE 6.11
Epithelial–mesenchymal and mesenchymal–epithelial transitions. (Modified from Weinberg, R.A., *The Biology of Cancer*, Garland Science, Taylor & Francis Group, New York, 2007.)

pylori, and *Vibrio cholera* have all been implicated in this process. *E. coli* (Nissle, 1917) increases intestinal epithelial permeability by decreasing the expression of zona occludens 2 (Zyrek et al., 2007), while *Bifidobacterium infantis* Y1 alters permeability by upregulating occludin and zona occludens 1 and downregulating claudin 2 expression (Ewaschuk et al., 2008). *H. pylori* and *V. cholera* both increase epithelial permeability by destroying tight junctions (Amieva et al., 2003; Krueger et al., 2007).

Excessive levels of reactive oxygen species and free radicals contribute to tight junction disruption, increased membrane permeability (Fraser, 2011), and blood–brain barrier destruction (Yamato et al., 2003). For example, xanthine oxidase generates reactive oxygen species in cultured endothelial cells (Beetsch et al., 1998), localizes to endothelial cells in the brain (Betz, 1985), and contributes to ischemic brain disease in animal experiments (Ono et al., 2009). Furthermore, incubation of rat brain endothelium with xanthine oxidase enables monocyte translocation across the blood–brain barrier via destruction of zona occludens 1 (Van der Goes et al., 2001) and relocation of occludin and zona occludens 1 (Lee et al., 2004).

Epithelial–mesenchymal transition is observed in the alveolar epithelium during idiopathic pulmonary fibrosis (Willis et al., 2005; Scotton and Chambers, 2007). The development of fibrosis is linked with the overexpression of transforming factor β1 in type II alveolar epithelial cells (Khalil et al., 1996). The type II epithelial cells undergo a transforming growth factor β1/ Smad protein–dependent epithelial–mesenchymal transition into myofibroblasts (Buckley et al., 2008; Kim et al., 2009; Shukla et al., 2009; Aoyagi-Ikeda et al., 2011; DeMaio et al., 2012). Epithelial–mesenchymal transition in idiopathic pulmonary fibrosis is also attributed to the downregulation of cadherins, tight junction proteins, and adhesion factors by free radicals (Gorowiec et al., 2012).

Similarly, epithelial–mesenchymal transition and the destruction of paracellular junctions play key roles in cancer metastasis (Takeichi, 1991, 1993; Brennan et al., 2010). For example, immunohistological examination of human metastatic breast cancer tissue showed a decrease in occludin expression with disease progression (Martin et al., 2010). Moreover, diminished expression of claudin 3 and claudin 4 was detected in stomach cancer (Jung et al., 2011), and claudin 7 expression was decreased in pulmonary cancer via activation of the extracellular-regulated kinase/mitogen-activated protein kinase cascade (Lu et al., 2011). Cancer cell metastasis entails endothelial–mesenchymal transition in addition to epithelial–mesenchymal transition. In this regard, claudin 5 directs the translocation of endothelial cells (Escudero-Esparza et al., 2012), and the downregulation of cadherin expression in endothelial cells is linked with tumor invasion and metastasis (Takeichi, 1991, 1993).

Occludin expression is markedly decreased in intestinal permeability disorders (e.g., Crohn's disease and ulcerative colitis), in keeping with the association between dysregulated occludin expression, increased permeability, and cancer cell metastasis. Indeed, occludin small-interfering RNA transfection increases transepithelial flux, even for dextrans as small as 7.2 nm (Al-Sadi et al., 2011).

Analysis of cellular mechanical properties (elasticity, hardness, and surface stiffness) in non–small cell lung cancer, breast ductal adenocarcinoma cells, and pancreatic adenocarcinoma cells showed that metastatic cancer cells (elastic modulus, 0.53 kPa) are 80% softer than normal cells (1.97 kPa). Prabhune et al. (2012) demonstrated that cancer cells are also softer than normal cells in other types of cancer, such as thyroid cancer. In addition, the surface adhesion values of cancer cells (34.2 pN) indicated that they are 33% less adhesive than normal cells (51.1 pN) (Cross et al., 2008).

The results with cancer cells imply that the use of soft versus hard nanoparticles might facilitate transport of nanomedicines. In support of this hypothesis, soft liposomes show high permeability after eye drop administration to the conjunctiva (see Chapter 5).

6.6 Practical Application of Nanomedicines

6.6.1 Nanoparticles in Disease

Injectable nanoproducts are engineered with targeting ability to avoid effects on normal cells while selectively delivering active ingredients to disease-associated cells. Conventional products have no such targeting ability. Conventional drugs of ≤5 nm pass though the pores of the capillary endothelium after intravenous injection, distribute to systemic tissues, and then are mainly eliminated by the kidney, partially secreted in the bile via the liver, or eliminated in the feces or through the dermal sweat glands.

The treatment regimen of an intravenously injected, conventional drug typically employs an excessive drug concentration so as to reach the target cells. After injection, the conventional drug is diluted by the volume of the blood and all the cells in the body, both normal and diseased. Therefore, the effective concentration depends on the total concentration diluted by the volume of the vasculature and the cellular volume. If a water tank model is used as a metaphor for the body, the total drug concentration is that of the water tank filled by cells and the effective concentration is that external to the cells. Because drugs in this model are delivered to all tissues and all cells in the body, the therapeutic and adverse effects are determined by the permeability of each cell to the drug.

By contrast, well-designed biodegradable nanomedicines of ≥5 nm are not immediately degraded in blood vessels and circulate throughout the body within the vasculature for a prolonged period of time. The restrictive blood vessel diameter ascertains that nanomedicines circulate at high concentrations. In the water tank model, the volume outside the cells and the vasculature is small. In agreement, pharmacokinetic studies in rats of 30 nm cisplatin-loaded, polymeric micelles showed that the volume of distribution was only ~1/75 that of free cisplatin (Uchino et al., 2005; Igarashi, 2008a). Therefore, circulating concentrations of intravenously injected nanomedicines are ~75-fold higher than those of conventional drugs.

Angiogenesis actively occurs around damaged or diseased tissue following external injury, infections, rheumatism, cerebrovascular accidents, and cancer. New blood capillaries are generated via angiogenesis, along with microscale paracellular gaps in the capillary endothelial wall. Angiogenesis during cancer temporarily creates endothelial gaps as large as 200–2000 nm

(Hobbs et al., 1998; Hashizume et al., 2000). Therefore, appropriately timed drug injections kill cancer cells.

Notably, nanomedicines effectively reach tumors and protect against metastasis due to the distinctive endothelial structure around cancer cells and the enhanced permeability and retention (EPR) effect (see Chapter 2) (Matsumura and Maeda, 1986; Maeda and Matsumura, 1989). The EPR effect is characterized by (1) leakiness of tumor-associated capillaries, (2) enhanced permeability of tumor cells to macromolecules of 15–70 kDa, and (3) low drug recovery through bloods and lymphatic vessels (Matsumura and Maeda, 1986; Maeda and Matsumura, 1989; Maeda et al., 2000; Igarashi, 2008a). The low recovery through lymphatic vessels results from the absence of a functional lymphatic system in tumor tissue (McDonald and Baluk, 2002; Igarashi, 2008a).

Anticancer agents include cytotoxic and noncytotoxic agents. Cytotoxic agents directly kill cancer cells by disrupting their life cycle or growth, while noncytotoxic agents interfere with cell transport or metabolism. Because cytotoxic anticancer agents are highly toxic, effective doses in humans frequently correspond to lethal doses in animals. The conventional cytotoxic agent, cisplatin, has a LD_{50} (median lethal dose) of 7–8 mg/kg (43–50 mg/m^2) in rats (Ito et al., 1981; Anabuki and Kohda, 1982), while the estimated maximal dose is 120 mg/m^2 in humans. However, conventional cisplatin for human use is used in continuous 24 h infusion regimens due to its lethality, implying that the delivered dose is two- to threefold higher than the rat LD_{50}. For this reason, nanoproducts loaded with cytotoxic anticancer drugs are under development to increase drug efficacy as well as safety. In the aforementioned discussion of 30 nm cisplatin-loaded, polymeric micelles, no death was observed at a dose equivalent to 15 mg/kg in the rat (Uchino et al., 2005), and the nanomedicine could safely be used at the maximal dose of 120 mg/m^2 cisplatin in cancer patients (Plummer et al., 2011). These findings show that nanomedicine platforms enable higher clinical doses in humans than conventional drugs.

Anticancer therapy generally comprises surgery, radiation therapy, and conventional chemotherapy. According to abundant cancer epidemiological data, this regimen is expected to be highly effective when the primary cancer is small in size and in the early phase of growth. However, when the cancer is large and in later phases of growth, the therapeutic effect diminishes.

Surgery, radiation therapy, and chemotherapy are all characterized by restrictions on their effectiveness. For example, the effectiveness of surgery is restricted when (1) the tumor is very small prior to metastasis; (2) metastatic cancer cells are discrete or invisible; (3) cancer cells or tissue are intermingled with normal cells; or (4) the location of the tumor renders surgery difficult or risky, such as in thyroid medullary cancer. Radiation therapy is restricted by the same concerns as those listed for surgery, in addition to the capacity to induce DNA mutations in normal and cancerous cells alike, increasing the risk of new cancer occurrence or cancer progression.

Conventional chemotherapeutic agents also provoke DNA mutations in normal and cancerous cells, and conventional cytotoxic agents kill both kinds of cell and cannot distinguish between the two.

6.6.2 Nanoproduct Engineering

Injectable nanomedicines are engineered for transport in the blood. Active ingredients are encapsulated within a shell composed of benign substances, such as amino acids, lipids, and/or glucose, resulting in a 10–200 nm nanomedicine. In this book, the delivery vehicle is referred to as the platform, while the encapsulated ingredients are referred to as the payload. Nanomedicines for injection include liposomes, polymeric micelles, albumin-binding nanoparticles, gold metallic nanoparticles, calcium phosphate nanoparticles, and virosomes.

LMW conventional drugs have only one molecular structure as the active ingredient, with the exception of excipients or component drugs. Conventional drugs must be able to interact with the membrane of target cells and escape rapid excretion via metabolism in the body. By contrast, nanomedicines usually contain a number of encapsulated ingredients in one capsule, including conventional drugs, substances with no pharmacological activity (e.g., RNA or proteins as building blocks for bottom-up assembly) by instable physicochemical characteristics or lacking the ability to entry into cell. The platform must directly interact with the membrane of target cells but not normal cells, escape rapid degradation by metabolism, permit uptake into target cells by endocytosis, and extend to nuclear translocation of drug components to the intended purpose of direct gene targeting. The active ingredient together with the excipients of conventional drugs must show all of these functions, and these functions are key for its efficacy and safety; these functions are largely shouldered by the platform in nanomedicines.

One platform can be used to produce countless nanomedicines. Countless nanomedicines can also be produced by varying the amount of active ingredient in each nanoparticle, as well as the amount and type of PEG coating used to prevent nonspecific interactions with cells other than the target cell and the amount and type of specific targeting ligand (protein, peptide, glucose, and so on). The efficacy and safety of a nanomedicine thus depends on the combination of platform, active ingredient, coating, and ligand. In comparisons with conventional drugs, the efficacy and toxicity of nanomedicine should be totally considered a nanoparticle.

Nanomedicines for intravenous injection in the oncology field include passive-targeting and active-targeting drugs. Passive-targeting nanomedicines are biodegradable nanoparticles coated with PEG to escape innate immunological responses, while active-targeting nanomedicines are nanoparticles coated with ligands that show high affinity and selectivity for the cancer cell surface.

Passive-targeting nanoparticles for oncology are degraded in the blood into fragments that include nutrient agents for cancer cells, such as amino acids or glucose. Many types of cancer cells preferentially uptake these nutrient agents via receptors (Ganapathy et al., 2009; Calvo et al., 2010; Broer and Palacin, 2011) or transporter-driven endocytosis, but passive-targeting nanoparticles are not intentionally designed with these receptors or transporters in mind. Rather, passive-targeting nanoparticles are defined as nontargeted, PEG-coated nanomedicines.

The basic engineering of nontargeted nanomedicines is geared toward the incorporation of stealth and self-biodegradability. The attribute of stealth allows intravenously injected nanoparticles to escape binding to enzymes or other proteins in the plasma. Nonstealthy nanoparticles can adversely affect both nonimmune and immune reactions. Nonimmune reactions include nanoparticle association with enzymes of the coagulation system, induction of thrombogenicity, decreased anticoagulant activity, and diminished detoxification capacity. Immune reactions include nanoparticle association with factors involved in the excretion of foreign substances and lowered immunity. Therefore, nanomedicines intended for long-term circulation in the blood are coated with PEG or another biocompatible substance to introduce stealth.

Nanoparticles engineered for biodegradability must be stable in liquid at the time of manufacture but degradable under appropriate physiological conditions. Various environmental parameters must be taken into account for the generation of a biodegradable nanoproduct, such as salt content in the blood, physiological pH and temperature, and oxygen concentration around cancer tissues. The main pharmacological effects depend on the actions of the degraded active ingredients and their restricted delivery to the target cell. The EPR effect allows optimal biodegradability in the vicinity of targeted cancer cells and should not be expected for nanoparticles with early or rapid biodegradability.

6.6.3 Albumin as a Carrier and Nanoproduct Stability in the Blood

Albumin-binding nanoparticles are produced by self-assembly of active nanomedicine ingredients and human albumin as the platform, exemplified by Abraxan®, an anticancer drug in the range of 50–150 nm (average, 130 nm). Abraxan is produced by the binding of paclitaxel to albumin and was approved by the U.S. FDA in 2005. Conventional paclitaxel is poorly soluble, creating pharmaceutical issues, while the addition of Cremophor, a synthetic nonionic surfactant, allows paclitaxel dissolution. However, Cremophor has numerous issues of its own, including precipitation of peripheral nerve/ganglion pain and hypersensitivity, as well as the disappearance of active drug.

Abraxan is a Cremophor-free formulation that is rapidly degraded within 30–45 s after injection, releasing paclitaxel and albumin into the blood.

Although Abraxan is a nanoparticle at the time of manufacture, the released paclitaxel shows the same disposition as conventional paclitaxel. Thus, Abraxan does not show selective accumulation in cancer tissue, a reduced frequency of adverse events by escaping normal tissue, or protective actions against metastatic cancer by the EPR effect. Therefore, Abraxan does not fit the definition of a passive-targeting product.

The albumin carrier of Abraxan is highly soluble, negatively charged, and monomeric. Albumin interacts with cysteine-rich acidic proteins, and the cationic albumin/protein complexes hypothetically interact with negatively charged cancer cell surfaces (Elsadek and Kratz, 2012). Albumin is also naturally occurring and has a long half-life of 19 days (Kratz, 2008), both of which are advantageous for its use as a drug carrier. Nonetheless, because Abraxan is efficacious as a suspension rather than as a water-soluble formulation, its preparation is not simple in clinical practice.

Albumin has also been applied as a drug carrier for the development of Albures® (diameter, 200–1000 nm) and Nacoll® (diameter, 8 nm), which are both ^{99m}Tc-labeled albumin nanoparticles for radiation diagnosis; Albuferon®, an interferon–albumin combination with a longer half-life than free interferon; Vasovist®, an albumin–gadolinium combination for magnetic resonance imaging (MRI) diagnosis of cancer; Herceptin® (trastuzumab), an albumin-bound, single-chain antibody (Fab4D5) directed against epidermal growth factor receptor 2; and Levemir® and Victoza®, both combinations of albumin with human insulin for diabetes therapy (Elsadek and Kratz, 2012).

Long-term stability of nanomedicines does not necessarily contribute to their stealth. For example, Doxil was engineered for stability based on its lipid constituents and for stealth based on its PEG coating, and the resulting liposome is indeed extremely stable in the blood. However, this formulation is almost nondegradable under physiological conditions, and liposome accumulation adversely affects cellular immunity. Bacteremia is a risk of Doxil injection in clinical practice because the liposome increases the immunological burden on macrophages and Kupffer cells and attenuates their ability to eliminate foreign substances (Zolnik and Sadrieh, 2009). The challenge of nanomedicine engineering is thus to generate nanoparticles with long-term stability and shelf life prior to use but appropriate time- and location-dependent self-biodegradability in the body, without involving metabolic enzymes or the innate immune system.

Active-targeting nanomedicines contain surface coatings of ligands against surface antigens (e.g., receptors) overexpressed in target cells, as well as encapsulated active ingredients for disease therapy. The basic engineering of an active-targeting nanomedicine for anticancer therapy comprises (1) selectivity against the cancer cell, (2) facilitation of uptake into the cancer cell, and (3) release of active ingredients inside the cell. Active-targeting nanomedicines have a large surface area compared with conventional drugs and are decorated with numerous ligand-based active sites, increasing the probability for interactions with and endocytosis by cancer cells.

Active-targeting nanomedicines enable therapy against small as well as large tumors and cancer cells interspersed among normal cells. The ligands are selected for their ability to target disease cells and to promote endocytosis, while the payload is selected to include both noncytotoxic and cytotoxic anticancer drugs and instable biologics in the blood (i.e., RNA or peptide-building blocks). The overall impact of an active-targeting nanomedicine depends on the type(s) of ligand, self-biodegradability, and releasability in the body or within the cell.

Active-targeting nanomedicines as described earlier employ Trojan horse tactics. The cancer cell recognizes and endocytoses the active-targeting nanomedicines due to the presentation of preferential substances by the latter to the cell. The cancer cell is then attacked from within by encapsulated active ingredients.

Recently marketed molecular-targeting drugs are antibody based and differ from active-targeting nanomedicines. Molecular-targeting drugs show no EPR effect and a limited interaction with target cells because only one active site provided by the antibody interacts with the target cell plasma membrane. Although one molecular-targeting drug, Gleevec®, showed high anticancer efficacy, other recently marketed molecular-targeting drugs showed no remarkable antitumor effect (Maeda and Matsumura, 2011).

Active-targeting nanomedicines are further discussed in Sections 6.7.4 through 6.7.6.

6.7 Engineering of Nanomedicines

Conventional drugs/biologics and polymeric micellar/liposomal nano-medicines show distinct biological and physical characteristics from one another (Table 6.2). Special characteristics of nanomedicines include stealth, self-biodegradability, constitution of the platform, the capacity of the payload to include multiple active ingredients, the high surface area of nanoparticles, ligand coatings, and the utilization of aptamers for selective gene targeting.

The engineering of nanoproducts intended for therapeutic use affects their safety and efficacy. The main engineering design includes (1) platform design in terms of size, shape, and rigidity; size and shape as a function of rigidity; and the amount of each active ingredient per nanoparticle; (2) the type and quantity of ligand coating for specific cell-targeting therapy; (3) the intracellular releasability of active ingredients from the nanoparticle after uptake by the cell; and (4) the coating of active ingredients with appropriate aptamers for nuclear translocation and direct gene targeting.

The cutoff size for opsonophagocytosis of particles by Kupffer cells in the liver is 200 nm (Harashima et al., 1994). Liposomes of ≤200 nm are not

TABLE 6.2

Biological and Physical Characteristics of Polymeric and Liposomal Nanomedicines and Conventional Drugs or Biologics

Characteristics	Drugs	Biologics	Passive-Targeting Nanomedicine (Polymer Micelle)	Active-Targeting Nanomedicine (Liposome)
Synthesis	Chemical synthesis[a]	Typically purified from genetically manipulated cell lines or isolated from biological fluid[a]	Incorporation of drug or biologic into a block copolymer[b]	Incorporation of drug or biologic into a liposome
Molecular weight	Low molecular weight[a]	High molecular weight[a]	High molecular weight[b]	High molecular weight
Particle size	Less than diameter 5 nm	Less than diameter 5 nm	5–100 nm in diameter	100–200 nm in diameter
Payload	Typically one active ingredient	Typically one active ingredient	Several ten active ingredients	Several hundred or thousand active ingredients
Uniformity of platform	Typically one molecule	Typically one molecule	Polydispersity	Polydispersity
Physical and chemical characteristics	Well-defined physical and chemical characteristics[a]	Complex physicochemical characteristics (i.e., tertiary structure, extent, and type of glycosylation)[a]	Polymer-dependent physicochemical characteristics (i.e., size, shape, and surface reactivity)[b]	Ligand-dependent physicochemical characteristics (i.e., size, shape, surface reactivity, rigidity)
Chemical stability	Generally stable and not particularly sensitive to heat[a]	Both heat and shear sensitive[a]	Generally stable against shear and sensitive to heat[b]	Generally stable against shear and sensitive to heat
Biological stability (biodegradability)	Metabolized[a]	Catabolized[a]	Autonomously or endogenously degraded[b]	Endogenously degraded, endocytosis by targeting cell[b]

Note: The descriptions of drugs and biologics as summarized by [a]Dempster (2000) were applied as well as the descriptions of passive-targeting nanomedicine summarized by [b]Igarashi (2008a) are modified so as to compare and contrast their physicochemical characteristics with those of nanomedicines.

[a] Indicates descriptions summarized by Dempster, 2000.
[b] Indicates descriptions summarized by Igarashi, 2008a.

efficiently taken up by opsonization into the Kupffer cell, indicating that injectable nanomedicines should be no larger than 200 nm.

6.7.1 Passive-Targeting Nanomedicines

Yokoyama et al. (1990, 1991) originally designed anionic polymeric nano-medicines produced by the self-assembly of self-degradable, 50 nm polymeric micelles in 1990. Since the first design of polymeric nanomedicines, many new formulations have appeared (Igarashi, 2008a; Matsumura, 2008). Polymeric micelles are an innovative nanoproduct that allows realization of the EPR effect. Polymeric micelles consist of encapsulated active ingredients as the payload, and PEG and polyamino acids (i.e., polyglutamine, polyasparagine, or polylysine) as the platform. These materials show a narrow polydispersity index at ≤100 nm. Self-assembly occurs via chemical binding or physical electrostatic interactions between PEG and the polyamino acid and between the polyamino acid and the active ingredients. Self-assembled nanomedicines are represented by polymeric micelles loaded with platinum-based drugs for cancer therapy and polymeric micelles loaded with gadolinium-based contrast agents for the MRI diagnosis of cancer (Yokoyama et al., 1996; Nishiyama et al., 1999; Nishiyama and Kataoka, 2001; Shiraishi et al., 2009, 2010).

6.7.2 Cancer Therapy

Platinum-loaded polymeric micelles for cancer therapy are made of anionic polymers and are coated with PEG for stealth. The polymeric micelles are formed through self-assembly of PEG and polyamino acids, as described earlier, and incorporate chemotherapy drugs that contain platinum metal derivatives (i.e., cisplatin). One such nanomedicine utilizes polymeric micelles of 20 or 28 nm and entraps cisplatin and polyasparagine (Yokoyama et al., 1996; Nishiyama et al., 1999; Nishiyama and Kataoka, 2001). The polymeric micelle degrades in the presence of salt and releases cisplatin in the vicinity of the tumor (Nishiyama et al., 2003). Another similar nanomedicine incorporates cisplatin and polyglutamine into 30 nm polymeric nanoparticles (Uchino et al., 2005; Igarashi, 2008a). These polymeric nanomedicines facilitate the EPR effect.

As described previously, a comparative pharmacokinetic study of the 30 nm cisplatin-loaded polymeric nanomedicine versus free cisplatin showed that the volume of distribution of the former was only 1/75 of the value for free cisplatin. Furthermore, the clearance of the nanomedicine was only 1/19 of the value for free cisplatin, corresponding to a 65-fold larger area under the curve (AUC) (Uchino et al., 2005). In addition, a phase I clinical study in cancer patients showed a 34-fold smaller C_{max} and a 8.5-fold larger AUC in the blood for a cisplatin dose of 120 mg/m^2 (Plummer et al., 2011). Therefore, cisplatin has efficacious anticancer activity in the form of platinum-loaded polymeric nanomedicines. For conventional anticancer chemotherapy, the applicable dose of free cisplatin is restricted by acute renal and peripheral nerve toxicity. However, the polymeric nanomedicines show lower renal and peripheral toxicity and 1.5-fold higher maximal tolerance.

6.7.3 Cancer Diagnosis after the Angiogenic Switch

Shiraishi et al. (2009) developed a gadolinium-loaded, PEG-coated polymeric micelle for the MRI-mediated diagnosis of cancer. The 43 nm polymeric micelle is formed by self-assembly of polylysine, PEG, and a gadolinium-based contrast agent and exhibits a neutral surface charge due to the chemical binding of cationic polylysine with the anionic gadolinium. The polymeric nanomedicine enabled the EPR effect in a pharmacokinetic study of rats and showed strong contrast around the site of cancer tissue at 24 h postinjection. Although any conventional gadolinium contrast agent can provoke acute renal toxicity by the release of gadolinium ions (Abraham and Thakral, 2008; Thakral and Abraham, 2009), the polymeric micelle was degraded in the blood and yielded degradation fragments in the urine that did not include free gadolinium ions (Shiraishi et al., 2010). The amelioration of acute renal toxicity was attributed to the lack of gadolinium ions due to the strong binding between polylysine and gadolinium.

Cancer begins with one transformed cell (Weinberg, 2007). Once clusters of approximately 10^4–10^5 transformed cells accumulate, the tumor begins to have a unique vascular structure and to show resistance against radiotherapy and chemotherapy (Frangioni, 2008) in a process termed the angiogenic switch. Conventional diagnostic means can only discern cancer once a cluster of ~10^9 tumor cells forms, corresponding to a size of 1 cm and a weight of 1 g. Cancer metastasis is dependent on tumor size, and ~22% of primary cancers have already progressed to the metastatic cancer stage in clusters as small as 0–1 cm (Weinberg, 2007). The new polymeric nanomedicines permit visualization of cancer clusters of ≤1 mm in experimental animals (Shiraishi et al., 2012) and are expected to assist in the detection of the onset of the angiogenic switch.

6.7.4 Active-Targeting Nanomedicines

Active-targeting nanomedicines for anticancer therapy are engineered to demonstrate selectivity against tumors due to the expression of specific cancer cell surface receptors and to have no selectivity against normal cells. However, cancer cells reflect the transdifferentiation of normal cells, and therefore cancer cell receptors are expressed in one form or another by a normal cell somewhere in the body. For example, the comprehensive analysis of membrane receptors in normal and transdifferentiated breast cancer cells showed that breast-cancer-specific receptors were also expressed on the normal cells of other organs. Because active-targeting nanomedicines expedite endocytosis of the drug into cancer cells and drug accumulation around tumors by the EPR effect, side effects to normal cells are minimized. Transferrin-targeted and octreotide-targeted liposomes are examples of active-targeting nanomedicines and are discussed in Sections 6.7.5 and 6.7.6.

6.7.5 Transferrin-Targeted Liposomes against Primary Cancer

Transferrin-targeted liposomes are lipid nanomedicines for therapy against primary cancer. The surface of nanoparticle is coated with transferrin as the ligand, and the corresponding transferrin receptor is overexpressed in cancer cells. Two types of transferrin receptor have been identified, type I and type II (Qian et al., 2002). Type I receptors show low expression in normal cells and high expression in cancer cells (Gatter et al., 1983), while type II receptors show high expression in normal liver and small intestine cells (Kawabata et al., 1999, 2000).

Kobayashi et al. (2007) showed that doxorubicin-loaded, transferrin-targeted liposomes are successfully endocytosed into cancer cells and released within the cell (Kobayashi et al., 2007). Moreover, Suzuki et al. (2008) conducted a pharmacokinetic study of oxaliplatin-loaded, transferrin-targeted 100 nm liposomes in mice bearing C-26 colon cancer cell–derived tumors. Injection of the targeted liposome showed a high distribution in cancer tissue but not in normal liver, and accumulation in cancer tissue was higher for the targeted liposome than for a liposome coated with PEG. In addition, intravenous injection of the transferrin-targeted liposome at 2.5 mg/kg oxaliplatin showed statistically significant improvements in tumor growth inhibition compared with free oxaliplatin at the same dose.

6.7.6 Octreotide-Targeted Liposomes

130-nm, octreotide-targeted liposomes are lipid nanoparticles with demonstrated efficacy against primary neuroendocrine cancer (Patel, 1999), metastatic liquid and solid cancers (Friedberg et al., 1999; Oomen et al., 2001), and stromal tissues neighboring tumors (Matei et al., 2012). The surface of the nanoparticle is coated with octreotide, an octapeptide analog of natural somatostatin. The platform is composed of octreotide, PEG, and lipids to generate the octreotide-poly(ethylene glycol)3400-distearoylphosphatidylethanolamine (PEG-DSPE) particle (Figure 6.12a).

Somatostatin receptors (SSTRs) belong to the GPCR family of transmembrane receptors and include SSTR1, SSTR2, SSTR3, SSTR4, and SSTR5. Octreotide binds to SSTR2, SSTR3, and SSTR5 with high affinity (Patel, 1999).

Iwase and Maitani (2011) studied cell uptake and drug releasability for irinotecan-loaded, octreotide-targeted liposomes in medullary thyroid

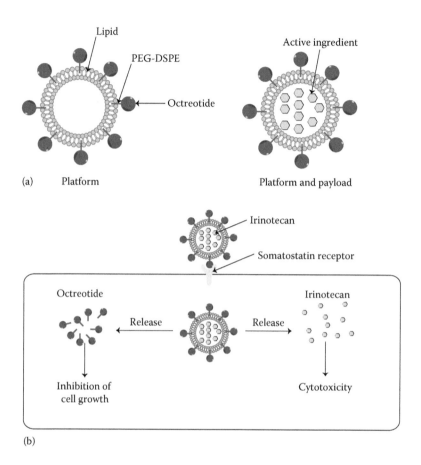

FIGURE 6.12
Schematic representation of the core formulation of octreotide-targeted liposomes and the anticancer mechanism of irinotecan-loaded, octreotide-targeted liposomes. (a) Formulation. (b) Mechanism of endocytosis by selective interaction, followed by intracellular release of irinotecan and octreotide. The dual effect of octreotide (growth inhibition) plus irinotecan (cytotoxicity) is illustrated. (Modified from Iwase, Y. and Maitani, Y., *Cancer Sci.*, 103(2), 310, 2012, doi: 10.1111/j.1349-7006.2011.02128.x.)

carcinoma cells. The liposome was endocytosed into the cells with subsequent release of irinotecan and conversion into the highly toxic SN-38 metabolite (Figure 6.12b). The liposome also exerted a stronger anticancer effect than free irinotecan in vivo. For instance, the liposome decreased tumor volume relative to baseline at 1/5 the dose of free irinotecan (Figure 6.13). Octreotide alone showed only growth inhibition of cancer cells with no cytotoxicity (Figure 6.12b). The high effect of the liposome formulation was attributed to octreotide in addition to irinotecan (Iwase and Maitani, 2012), and the empty octreotide liposome was actually more effective against tumor cell growth than free octreotide in vitro.

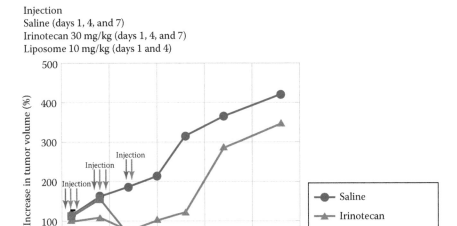

Injection
Saline (days 1, 4, and 7)
Irinotecan 30 mg/kg (days 1, 4, and 7)
Liposome 10 mg/kg (days 1 and 4)

FIGURE 6.13
Antitumor effect of octreotide-based active-targeting liposome. Square, octreotide-based active-targeting liposome; triangle, free irinotecan; circle, saline. (Modified from Iwase, Y. and Maitani, Y., *Mol. Pharmaceut.*, 8(2), 330, 2011, doi: 10.1021/Mp100380y.)

Cancer cell growth and proliferation are associated with the phosphorylation of p70S6K, where phosphorylated p70S6K serves as a marker of the inhibition of the PI3K/Akt/mTOR signaling pathway. Treatment with the empty octreotide liposome at a low dose of 0.42 µM suppressed phosphorylation of p70S6K in vitro and in vivo, whereas a plain (no octreotide) empty liposome or free octreotide at a high dose of 100 µM did not.

Embryogenesis and carcinogenesis are closely related processes (Sabe, 2011; Brellier and Chiquet-Ehrismann, 2012). Embryogenesis involves the differentiation of a single fertilized ovum into the three germ layers (the endoderm, ectoderm, and mesoderm) and the subsequent differentiation of germ layer cells into all the cells of the adult body. For example, melanocyte differentiation comprises delamination of neural crest cells from the neuroectoderm, migration throughout the embryo, and differentiation into melanoblasts and then melanocytes. Carcinogenesis involves the transdifferentiation of mature cells into undifferentiated cells of the same origin via the same pathway or the dedifferentiation of mature cells into undifferentiated cells of another origin via a different pathway (Eisenberg and Eisenberg, 2003). The loss of intercellular adhesions in metastatic cancer is similar to that of migrating cells during embryogenesis (Sabe, 2011; Brellier and Chiquet-Ehrismann, 2012).

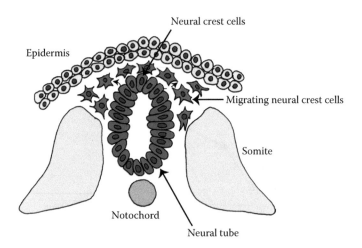

FIGURE 6.14
Schematic representation of SSTR expression during the migration of neural crest cells. Migrating neural crest cells lose cell–cell communication ability and migrate to the mesodermal space. The loss of cell–cell communication results in a phenotype reminiscent of the metastatic tumor cell.

SSTRs are highly expressed in embryogenesis and carcinogenesis. Neural crest cells show elevated expression of somatostatin and catecholamine during the migratory phase after neural tube closure in early embryogenesis (Figure 6.14), but not during the later phases of embryogenesis (Maxwell and Sietz, 1985; Garcia-Arraras and Torres-Avillan, 1999), while SSTR expression almost disappears after birth, except for in carcinogenesis. Somatostatin and its octreotide analog have no binding affinity for normal adult human liver cells, but they show high binding affinity for hepatocarcinoma cells.

The affinity of octreotide for cancer tissue is associated with the location of the cancer and the type of SSTR expressed. According to gene analysis, SSTRs comprise two groups, one consisting of SSTR1 and SSTR4 subtypes, and the other consisting of SSTR2, SSTR3, and SSTR5 subtypes (Figure 6.15) (Hannon et al., 2002). As noted earlier, octreotide binds with high affinity to SSTR2, SSTR3, and SSTR5 (Patel, 1999), but not to SSTR1 and SSTR4. SSTR1 and SSTR4 are overexpressed in prostate cancer and colon cancer, while SSTR2, SSTR3, and SSTR5 are overexpressed in neuroendocrine, melanoma, GI, pancreatic, liver, renal, small cell lung cancer, brain, and breast cancer cells (Lum et al., 2001; Papotti et al., 2001; Reubi et al., 2001; Blaker et al., 2004; Orlando et al., 2004). When primary cancer changes to metastatic cancer, SSTR2 in particular is overexpressed in "liquid tumors," or blood cancers such as leukemia, as well as in solid sarcomas (Friedberg et al., 1999; Oomen et al., 2001). Recently, SSTR2 was also shown to be upregulated in stromal tissue adjacent to prostate cancer (Matei et al., 2012).

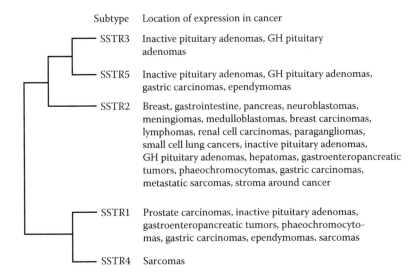

Subtype | Location of expression in cancer
SSTR3 | Inactive pituitary adenomas, GH pituitary adenomas
SSTR5 | Inactive pituitary adenomas, GH pituitary adenomas, gastric carcinomas, ependymomas
SSTR2 | Breast, gastrointestine, pancreas, neuroblastomas, meningiomas, medulloblastomas, breast carcinomas, lymphomas, renal cell carcinomas, paragangliomas, small cell lung cancers, inactive pituitary adenomas, GH pituitary adenomas, hepatomas, gastroenteropancreatic tumors, phaeochromocytomas, gastric carcinomas, metastatic sarcomas, stroma around cancer
SSTR1 | Prostate carcinomas, inactive pituitary adenomas, gastroenteropancreatic tumors, phaeochromocytomas, gastric carcinomas, ependymomas, sarcomas
SSTR4 | Sarcomas

FIGURE 6.15
Relationship between SSTR subtype and cancer. (Modified from Hannon, J.P. et al., *J. Mol. Neurosci.*, 18(1–2), 15, 2002.)

Taken together, these findings imply that octreotide-targeted liposomes show general selectivity against cancer by the EPR effect due to the behavior of the liposome as a nanoparticle. The octreotide ligand confers selectivity against primary cancer at the tissue level according to the SSTR subtype (SSTR2, SSTR3, SSTR5), against liquid and solid metastatic cancer according to the presence of SSTR2, and against SSTR2-expressing stromal tissue neighboring tumors.

6.8 Engineering of Nanoproducts Targeting Disease

The prolonged circulation of nanoproducts in the blood leads to the accumulation of stable products in areas of external injury, infection, disease, rheumatism, and cancer. In cancer therapy, stealthy, stable nanomedicines heighten therapeutic potential and improve drug safety. The benefits depend on cancer type, particle size, biodegradability, and targeting.

The intercellular space of the endothelium around cancer tissue ranges from 200 to 2000 nm. Therefore, nanoparticles of ≤100 nm as well as those of ~150 nm accumulate around cancer tissue by the EPR effect and are efficacious drug delivery agents. The gaps and pores in the endothelium adjacent to cancer tissue are variable in size and depend on the location and type of cancer (Hobbs et al., 1998; Hashizume et al., 2000; Morikawa et al., 2002; Igarashi, 2008a). It is not clear whether nanoparticles of ≤100 nm have an advantage compared with those of 150 nm in terms of endothelial permeability.

Cancer cells are of two types and are either (1) large in size and clump forming or (2) small in size and sporadically appearing amidst normal cells. Both normal cells and cancerous cells can endocytose nanoparticles of 200 nm or less via the clathrin pathway (Rejman et al., 2004) and, therefore, both normal and cancerous cells can endocytose passive-targeting nanomedicines. Active-targeting nanomedicines show selectivity against cancer cells. Because cancer cells can also endocytose active-targeting nanomedicines by high affinity, active-targeting nanoparticles minimize the probability of drug endocytosis by normal cells. The interaction of active-targeting nanoparticles with cancer cells via ligand–receptor binding also gives active-targeting nanoparticles an advantage by prolonging the time of association with the cancer cell.

6.9 Nanoscale World via the Systemic Route

The cutoff size for nanoproduct transport through the endothelial/epithelial transcellular and paracellular pathways of the systemic route, as well as endocytosis into target cells, is described in this chapter and illustrated in Figure 6.16.

The cutoff size for transport through the pores in the renal endothelium (the main pathway for excretion) is ~5 nm. The liver sinusoid contains pores and gaps of various sizes (≤100, 100–500 and ≥500 nm, including gaps of 1–3 µm). Nanoproducts can enter the Disse space of the liver through capillaries, but hepatic tissue can control the opening and closing of capillary pores in response to environmental conditions. Nanoparticles of ≥200 nm are endocytosed by Kupffer cells lining the wall of the liver sinusoid. Therefore, the size of injectable nanoproducts is limited to 200 nm or less.

The cutoff size of the transcellular pore of the continuous-type endothelium is ≤1 nm, while the cutoff size for paracellular gaps near cancer tissue is ~2 µm. Despite the dense ultrastructure of the laminin/collagen IV–based basal lamina, nanosized pores are found in the basement membranes of various tissues. Basement membrane pores of 13–222 and 11–30 nm are found in the bladder and renal glomeruli, respectively. The basement membrane of the human bronchial airway contains pores of 0.6–4 µm, thus enabling easy passage of microparticles.

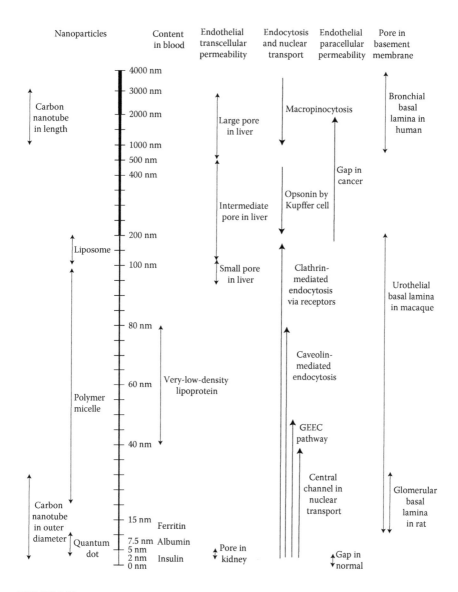

FIGURE 6.16
The nanoscale world via the systemic route.

References

Abraham, J. L. and C. Thakral. 2008. Tissue distribution and kinetics of gadolinium and nephrogenic systemic fibrosis. *European Journal of Radiology* 66 (2):200–207. doi: 10.1016/j.ejrad.2008.01.026.

Abrams, G. A., S. L. Goodman, P. F. Nealey, M. Franco, and C. J. Murphy. 2000a. Nanoscale topography of the basement membrane underlying the corneal epithelium of the rhesus macaque. *Cell and Tissue Research* 299 (1):39–46.

Abrams, G. A., C. J. Murphy, Z. Y. Wang, P. F. Nealey, and D. E. Bjorling. 2003. Ultrastructural basement membrane topography of the bladder epithelium. *Urological Research* 31 (5):341–346. doi: 10.1007/s00240-003-0347-9.

Abrams, G. A., S. S. Schaus, S. L. Goodman, P. F. Nealey, and C. J. Murphy. 2000b. Nanoscale topography of the corneal epithelial basement membrane and Descemet's membrane of the human. *Cornea* 19 (1):57–64.

Aggarwal, P., J. B. Hall, C. B. McLeland, M. A. Dobrovolskaia, and S. E. McNeil. 2009. Nanoparticle interaction with plasma proteins as it relates to particle biodistribution, biocompatibility and therapeutic efficacy. *Advanced Drug Delivery Reviews* 61 (6):428–437. doi: 10.1016/j.addr.2009.03.009.

Aisen, P., C. Enns, and M. Wessling-Resnick. 2001. Chemistry and biology of eukaryotic iron metabolism. *International Journal of Biochemistry & Cell Biology* 33 (10): 940–959.

Al-Sadi, R., K. Khatib, S. H. Guo, D. M. Ye, M. Youssef, and T. Ma. 2011. Occludin regulates macromolecule flux across the intestinal epithelial tight junction barrier. *American Journal of Physiology—Gastrointestinal and Liver Physiology* 300 (6): G1054–G1064. doi: 10.1152/ajpgi.00055.2011.

Amieva, M. R., R. Vogelmann, A. Covacci, L. S. Tompkins, W. J. Nelson, and S. Falkow. 2003. Disruption of the epithelial apical-junctional complex by *Helicobacter pylori* CagA. *Science* 300 (5624):1430–1434. doi: 10.1126/science.1081919.

Anabuki, K. and S. Kohda. 1982. Cisplatin: Acute toxicity in mice and rats. *Japanese Pharmacology and Therapeutics (Japanese)* 10:181–189.

Anderegg, J. W., W. W. Beeman, S. Shulman, and P. Kaesberg. 1955. An investigation of the size, shape and hydration of serum albumin by small-angle X-ray scattering. *Journal of the American Chemical Society* 77 (11):2927–2937.

Angelow, S., R. Ahlstrom, and A. S. L. Yu. 2008. Biology of claudins. *American Journal of Physiology—Renal Physiology* 295 (4):F867–F876. doi: 10.1152/ajprenal.90264.2008.

Aoyagi-Ikeda, K., T. Maeno, H. Matsui, M. Ueno, K. Hara, Y. Aoki, F. Aoki et al. 2011. Notch induces myofibroblast differentiation of alveolar epithelial cells via transforming growth factor-β-Smad3 pathway. *American Journal of Respiratory Cell and Molecular Biology* 45 (1):136–144. doi: 10.1165/rcmb.2010-0140OC.

Avasthi, P. S. and V. Koshy. 1988a. The anionic matrix at the rat glomerular endothelial surface. *Anatomical Record* 220 (3):258–266.

Avasthi, P. S. and V. Koshy. 1988b. Glomerular endothelial glycocalyx. *Contributions to Nephrology* 68:104–113.

Ballermann, B. J. 2005. Diaphragmed fenestrae in the glomerular endothelium versus nondiaphragmed fenestrae in the hepatic endothelium—Reply from the authors. *Kidney International* 68 (4):1902–1903.

Ballermann, B. J. and R. V. Stan. 2007. Resolved: Capillary endothelium is a major contributor to the glomerular filtration barrier. *Journal of the American Society of Nephrology (JASN)* 18 (9):2432–2438. doi: 10.1681/ASN.2007060687.

Bearer, E. L., L. Orci, and P. Sors. 1985. Endothelial fenestral diaphragms—A quick-freeze, deep-etch study. *Journal of Cell Biology* 100 (2):418–428. doi: 10.1083/jcb.100.2.418.

Beetsch, J. W., T. S. Park, L. L. Dugan, A. R. Shah, and J. M. Gidday. 1998. Xanthine oxidase-derived superoxide causes reoxygenation injury of ischemic cerebral endothelial cells. *Brain Research* 786 (1–2):89–95. doi: 10.1016/s0006-8993(97)01407-8.

Betz, A. L. 1985. Identification of hypoxanthine transport and xanthine-oxidase activity in brain capillaries. *Journal of Neurochemistry* 44 (2):574–579. doi: 10.1111/j.1471-4159.1985.tb05451.x.

Blaker, M., M. Schmitz, A. Gocht, S. Burghardt, M. Schulz, D. C. Broring, A. Pace, H. Greten, and A. de Weerth. 2004. Differential expression of somatostatin receptor subtypes in hepatocellular carcinomas. *Journal of Hepatology* 41 (1):112–118. doi: 10.1016/j.hep.2004.03.018.

Braet, F., R. Dezanger, M. Baekeland, E. Crabbe, P. Vandersmissen, and E. Wisse. 1995. Structure and dynamics of the fenestrae-associated cytoskeleton of rat-liver sinusoidal endothelial-cells. *Hepatology* 21 (1):180–189.

Brellier, F. and R. Chiquet-Ehrismann. 2012. How do tenascins influence the birth and life of a malignant cell? *Journal of Cellular and Molecular Medicine* 16 (1):32–40. doi: 10.1111/j.1582-4934.2011.01360.x.

Brennan, K., G. Offiah, E. A. McSherry, and A. M. Hopkins. 2010. Tight junctions: A barrier to the initiation and progression of breast cancer? *Journal of Biomedicine and Biotechnology* 2010:460607. doi: 10.1155/2010/460607.

Broer, S. and M. Palacin. 2011. The role of amino acid transporters in inherited and acquired diseases. *Biochemical Journal* 436:193–211. doi: 10.1042/Bj20101912.

Buckley, S., W. Shi, L. Barsky, and D. Warburton. 2008. TGF-β signaling promotes survival and repair in rat alveolar epithelial type 2 cells during recovery after hyperoxic injury. *American Journal of Physiology—Lung Cellular and Molecular Physiology* 294 (4):L739–L748. doi: 10.1152/ajplung.00294.2007.

Calvo, M. B., A. Figueroa, E. G. Pulido, R. G. Campelo, and L. A. Aparicio. 2010. Potential role of sugar transporters in cancer and their relationship with anti-cancer therapy. *International Journal of Endocrinology* 2010:205357. doi: 10.1155/2010/205357.

Casals, E., T. Pfaller, A. Duschl, G. J. Oostingh, and V. Puntes. 2010. Time evolution of the nanoparticle protein corona. *ACS Nano* 4 (7):3623–3632. doi: 10.1021/Nn901372t.

Chanan-Khan, A., J. Szebeni, S. Savay, L. Liebes, N. M. Rafique, C. R. Alving, and F. M. Muggia. 2003. Complement activation following first exposure to pegylated liposomal doxorubicin (Doxil): Possible role in hypersensitivity reactions. *Annals of Oncology* 14 (9):1430–1437. doi: 10.1093/annonc/mdg374.

Choi, H. S., W. Liu, P. Misra, E. Tanaka, J. P. Zimmer, B. I. Ipe, M. G. Bawendi, and J. V. Frangioni. 2007. Renal clearance of quantum dots. *Nature Biotechnology* 25 (10):1165–1170. doi: 10.1038/Nbtl340.

Cohen, S., S. Au, and N. Pante. 2011. How viruses access the nucleus. *Biochimica et Biophysica Acta—Molecular Cell Research* 1813 (9):1634–1645. doi: 10.1016/j.bbamcr.2010.12.009.

Colognato, H., D. A. Winkelmann, and P. D. Yurchenco. 1999. Laminin polymerization induces a receptor-cytoskeleton network. *Journal of Cell Biology* 145 (3): 619–631. doi: 10.1083/jcb.145.3.619.

Cross, S. E., Y. S. Jin, J. Tondre, R. Wong, J. Rao, and J. K. Gimzewski. 2008. AFM-based analysis of human metastatic cancer cells. *Nanotechnology* 19 (38):384003. doi: 10.1088/0957-4484/19/38/384003.

Deli, M. A. 2009. Potential use of tight junction modulators to reversibly open membranous barriers and improve drug delivery. *Biochimica et Biophysica Acta* 1788 (4): 892–910. doi: 10.1016/j.bbamem.2008.09.016.

DeMaio, L., S. T. Buckley, M. S. Krishnaveni, P. Flodby, M. Dubourd, A. Banfalvi, Y. M. Xing et al. 2012. Ligand-independent transforming growth factor-ss type I receptor signalling mediates type I collagen-induced epithelial-mesenchymal transition. *Journal of Pathology* 226 (4):633–644. doi: 10.1002/Path.3016.

Dempster, A. M. 2000. Nonclinical safety evaluation of biotechnologically derived pharmaceuticals. *Biotechnology Annual Review* 5:221–258.

Dickerman, H. W. and W. G. Walker. 1964. Effect of cationic amino acid infusion on potassium metabolism in vivo. *American Journal of Physiology* 206 (2):403–408.

Dobrovolskaia, M. A., P. Aggarwal, J. B. Hall, and S. E. McNeil. 2008. Preclinical studies to understand nanoparticle interaction with the immune system and its potential effects on nanoparticle biodistribution. *Molecular Pharmaceutics* 5 (4):487–495. doi: 10.1021/Mp800032f.

Dobrovolskaia, M. A., A. K. Patri, J. W. Zheng, J. D. Clogston, N. Ayub, P. Aggarwal, B. W. Neun, J. B. Hall, and S. E. McNeil. 2009. Interaction of colloidal gold nanoparticles with human blood: Effects on particle size and analysis of plasma protein binding profiles. *Nanomedicine—Nanotechnology Biology and Medicine* 5 (2):106–117. doi: 10.1016/j.nano.2008.08.001.

Durbeej, M. 2010. Laminins. *Cell and Tissue Research* 339 (1):259–268. doi: 10.1007/s00441-009-0838-2.

Ebnet, K. 2008. Organization of multiprotein complexes at cell-cell junctions. *Histochemistry and Cell Biology* 130 (1):1–20. doi: 10.1007/s00418-008-0418-7.

Edinger, T. O., M. O. Pohl, and S. Stertz. 2014. Entry of influenza A virus: Host factors and antiviral targets. *Journal of General Virology* 95:263–277. doi: 10.1099/Vir.0.059477-0.

Eisenberg, L. M. and C. A. Eisenberg. 2003. Stem cell plasticity, cell fusion, and transdifferentiation. *Birth Defects Research. Part C, Embryo Today: Reviews* 69 (3):209–218. doi: 10.1002/bdrc.10017.

Ekblom, P., P. Lonai, and J. F. Talts. 2003. Expression and biological role of laminin-1. *Matrix Biology* 22 (1):35–47. doi: 10.1016/s0945-053x(03)00015-5.

Elsadek, B. and F. Kratz. 2012. Impact of albumin on drug delivery—New applications on the horizon. *Journal of Controlled Release* 157 (1):4–28. doi: 10.1016/j.jconrel.2011.09.069.

Escudero-Esparza, A., W. G. Jiang, and T. A. Martin. 2012. Claudin-5 participates in the regulation of endothelial cell motility. *Molecular and Cellular Biochemistry* 362 (1–2):71–85. doi: 10.1007/s11010-011-1129-2.

Ewaschuk, J. B., H. Diaz, L. Meddings, B. Diederichs, A. Dmytrash, J. Backer, M. L. V. Langen, and K. L. Madsen. 2008. Secreted bioactive factors from *Bifidobacterium infantis* enhance epithelial cell barrier function. *American Journal of Physiology—Gastrointestinal and Liver Physiology* 295 (5):G1025–G1034. doi: 10.1152/ajpgi.90227.2008.

Favoreel, H. W., L. W. Enquist, and B. Feierbach. 2007. Actin and Rho GTPases in herpesvirus biology. *Trends in Microbiology* 15 (9):426–433. doi: 10.1016/j.tim. 2007.08.003.

Frangioni, J. V. 2008. New technologies for human cancer imaging. *Journal of Clinical Oncology* 26 (24):4012–4021. doi: 10.1200/Jco.2007.14.3065.

Fraser, P. A. 2011. The role of free radical generation in increasing cerebrovascular permeability. *Free Radical Biology and Medicine* 51 (5):967–977. doi: 10.1016/j.freeradbiomed.2011.06.003.

Friedberg, J. W., A. D. Van den Abbeele, K. Kehoe, S. Singer, C. D. Fletcher, and G. D. Demetri. 1999. Uptake of radiolabeled somatostatin analog is detectable in patients with metastatic foci of sarcoma. *Cancer* 86 (8):1621–1627.

Friedrich, B. M., N. Dziuba, G. Y. Li, M. A. Endsley, J. L. Murray, and M. R. Ferguson. 2011. Host factors mediating HIV-1 replication. *Virus Research* 161 (2):101–114. doi: 10.1016/j.virusres.2011.08.001.

Frise, A. E., E. Edri, I. Furo, and O. Regev. 2010. Protein dispersant binding on nanotubes studied by NMR self-diffusion and cryo-TEM techniques. *Journal of Physical Chemistry Letters* 1 (9):1414–1419. doi: 10.1021/jz100342c.

Fujita, H. and T. Fujita. 1992. *Textbook of Histology: Part 2*, 3rd edn. Tokyo, Japan: Igaku-Shoin Ltd.

Fujita, T., J. Tokunaga, and M. Miyoshi. 1970. Scanning electron microscopy of the podocytes of renal glomerulus. *Archivum Histologicum Japonicum* 32 (2): 99–113.

Furuse, M., K. Fujita, T. Hiiragi, K. Fujimoto, and S. Tsukita. 1998a. Claudin-1 and -2: Novel integral membrane proteins localizing at tight junctions with no sequence similarity to occludin. *Journal of Cell Biology* 141 (7):1539–1550.

Furuse, M., T. Hirase, M. Itoh, A. Nagafuchi, S. Yonemura, S. Tsukita, and S. Tsukita. 1993. Occludin—A novel integral membrane-protein localizing at tight junctions. *Journal of Cell Biology* 123 (6):1777–1788.

Furuse, M., H. Sasaki, K. Fujimoto, and S. Tsukita. 1998b. A single gene product, claudin-1 or -2, reconstitutes tight junction strands and recruits occludin in fibroblasts. *Journal of Cell Biology* 143 (2):391–401.

Ganapathy, V., M. Thangaraju, and P. D. Prasad. 2009. Nutrient transporters in cancer: Relevance to Warburg hypothesis and beyond. *Pharmacology & Therapeutics* 121 (1):29–40. doi: 10.1016/j.pharmthera.2008.09.005.

Garcia-Arraras, J. E. and I. Torres-Avillan. 1999. Developmental expression of galanin-like immunoreactivity by members of the avian sympathoadrenal cell lineage. *Cell and Tissue Research* 295 (1):33–41.

Gatter, K. C., G. Brown, I. S. Trowbridge, R. E. Woolston, and D. Y. Mason. 1983. Transferrin-receptors in human-tissues-their distribution and possible clinical relevance. *Journal of Clinical Pathology* 36 (5):539–545. doi: 10.1136/jcp.36.5.539.

Gloriam, D. E., R. Fredriksson, and H. B. Schioth. 2007. The G protein-coupled receptor subset of the rat genome. *BMC Genomics* 8:338. doi: 10.1186/1471-2164-8-338.

Gorowiec, M. R., L. A. Borthwick, S. M. Parker, J. A. Kirby, G. C. Saretzki, and A. J. Fisher. 2012. Free radical generation induces epithelial-to-mesenchymal transition in lung epithelium via a TGF-β1-dependent mechanism. *Free Radical Biology and Medicine* 52 (6):1024–1032. doi: 10.1016/j.freeradbiomed.2011.12.020.

Gu, Y. J., J. P. Cheng, C. C. Lin, Y. W. Lam, S. H. Cheng, and W. T. Wong. 2009. Nuclear penetration of surface functionalized gold nanoparticles. *Toxicology and Applied Pharmacology* 237 (2):196–204. doi: 10.1016/j.taap.2009.03.009.

Hallmann, R., N. Horn, M. Selg, O. Wendler, F. Pausch, and L. M. Sorokin. 2005. Expression and function of laminins in the embryonic and mature vasculature. *Physiological Reviews* 85 (3):979–1000.

Hannon, J. P., C. Nunn, B. Stolz, C. Bruns, G. Weckbecker, I. Lewis, T. Troxler, K. Hurth, and D. Hoyer. 2002. Drug design at peptide receptors—Somatostatin receptor ligands. *Journal of Molecular Neuroscience* 18 (1–2):15–27.

Haraldsson, B., J. Nystrom, and W. M. Deen. 2008. Properties of the glomerular barrier and mechanisms of proteinuria. *Physiological Reviews* 88 (2):451–487. doi: 10.1152/physrev.00055.2006.

Harashima, H., K. Sakata, K. Funato, and H. Kiwada. 1994. Enhanced hepatic-uptake of liposomes through complement activation depending on the size of liposomes. *Pharmaceutical Research* 11 (3):402–406. doi: 10.1023/A:1018965121222.

Harrison, P. M. and P. Arosio. 1996. Ferritins: Molecular properties, iron storage function and cellular regulation. *Biochimica et Biophysica Acta—Bioenergetics* 1275 (3):161–203. doi: 10.1016/0005-2728(96)00022-9.

Hashizume, H., P. Baluk, S. Morikawa, J. W. McLean, G. Thurston, S. Roberge, R. K. Jain, and D. M. McDonald. 2000. Openings between defective endothelial cells explain tumor vessel leakiness. *American Journal of Pathology* 156:1363–1380.

Hirano, S. and M. Takeichi. 2012. Cadherins in brain morphogenesis and wiring. *Physiological Reviews* 92 (2):597–634. doi: 10.1152/physrev.00014.2011.

Hironaka, K., H. Makino, Y. Yamasaki, and Z. Ota. 1993. Renal basement-membranes by ultrahigh resolution scanning electron-microscopy. *Kidney International* 43 (2):334–345.

Hobbs, S. K., W. L. Monsky, F. Yuan, W. G. Roberts, L. Griffith, V. P. Torchilin, and R. K. Jain. 1998. Regulation of transport pathways in tumor vessels: Role of tumor type and microenvironment. *Proceedings of the National Academy of Sciences of the United States of America* 95:4607–4612.

Hoelz, A., E. W. Debler, and G. Blobel. 2011. The structure of the nuclear pore complex. *Annual Review of Biochemistry* 80:613–643. doi: 10.1146/annurev-biochem-060109-151030.

Horn, T., J. H. Henriksen, and P. Christoffersen. 1986. The sinusoidal lining cells in normal human liver. A scanning electron microscopic investigation. *Liver International* 6:98–110.

Howat, W. J., J. A. Holmes, S. T. Holgate, and P. M. Lackie. 2001. Basement membrane pores in human bronchial epithelium—A conduit for infiltrating cells? *American Journal of Pathology* 158 (2):673–680. doi: 10.1016/S0002-9440(10)64009-6.

Igarashi, E. 2008a. Factors affecting toxicity and efficacy of polymeric nanomedicines. *Toxicology and Applied Pharmacology* 229 (1):121–134. doi: 10.1016/j.taap.2008.02.007.

Igarashi, E. 2008b. Factors affecting toxicity and efficacy of injectable nanomedicines: Significance on particle size. In: *Paper Read at EHRLICH II—Second World Conference on Magic Bullets, Celebrating the 100th Anniversary of the Nobel Prize Award to Paul Ehrlich*, May 15, Nürnberg, Germany.

Ikenouchi, J., M. Furuse, K. Furuse, H. Sasaki, S. Tsukita, and S. Tsukita. 2005. Tricellulin constitutes a novel barrier at tricellular contacts of epithelial cells. *Journal of Cell Biology* 171 (6):939–945. doi: 10.1083/jcb.200510043.

Ishimura, K., H. Okamoto, and H. Fujita. 1978. Freeze-etching images of capillary endothelial pores in liver, thyroid and adrenal of mouse. *Archivum Histologicum Japonicum* 41 (2):187–193.

Ito, K., T. Sakakibara, J. Hanta, Y. Irie, M. Tsubosaki, and A. Matsuda. 1981. Safety assessment study of cisplatin (NK801) (First report): Acute toxicity study in mice, rats and dogs. *Clinical Reports (Japanese)* 15:5–22.

Itoh, M., M. Furuse, K. Morita, K. Kubota, M. Saitou, and S. Tsukita. 1999. Direct binding of three tight junction-associated MAGUKs, ZO-1, ZO-2 and ZO-3, with the COOH termini of claudins. *Journal of Cell Biology* 147 (6):1351–1363.

Itoh, M., A. Nagafuchi, S. Yonemura, T. Kitaniyasuda, S. Tsukita, and S. Tsukita. 1993. The 220-Kd protein colocalizing with cadherins in nonepithelial cells is identical to Zo-1, a tight junction associated protein in epithelial-cells—cDNA cloning and immunoelectron microscopy. *Journal of Cell Biology* 121 (3):491–502.

Iwase, Y. and Y. Maitani. 2011. Octreotide-targeted liposomes loaded with CPT-11 enhanced cytotoxicity for the treatment of medullary thyroid carcinoma. *Molecular Pharmaceutics* 8 (2):330–337. doi: 10.1021/Mp100380y.

Iwase, Y. and Y. Maitani. 2012. Dual functional octreotide-modified liposomal irinotecan leads to high therapeutic efficacy for medullary thyroid carcinoma xenografts. *Cancer Science* 103 (2):310–316. doi: 10.1111/j.1349-7006.2011.02128.x.

Jung, H., K. H. Jun, J. H. Jung, H. M. Chin, and W. B. Park. 2011. The expression of claudin-1, claudin-2, claudin-3, and claudin-4 in gastric cancer tissue. *Journal of Surgical Research* 167 (2):E185–E191. doi: 10.1016/j.jss.2010.02.010.

Kawabata, H., R. S. Germain, P. T. Vuong, T. Nakamaki, J. W. Said, and H. P. Koeffler. 2000. Transferrin receptor 2-α; supports cell growth both in iron-chelated cultured cells and in vivo. *Journal of Biological Chemistry* 275 (22):16618–16625. doi: 10.1074/jbc.M908846199.

Kawabata, H., S. Yang, T. Hirama, P. T. Vuong, S. Kawano, A. F. Gombart, and H. P. Koeffler. 1999. Molecular cloning of transferrin receptor 2—A new member of the transferrin receptor-like family. *Journal of Biological Chemistry* 274 (30):20826–20832. doi: 10.1074/jbc.274.30.20826.

Khalil, N., R. N. O'Connor, K. C. Flanders, and H. Unruh. 1996. TGF-β(1), but not TGF-β(2) or TGF-β(3), is differentially present in epithelial cells of advanced pulmonary fibrosis: An immunohistochemical study. *American Journal of Respiratory Cell and Molecular Biology* 14 (2):131–138.

Kim, K. K., Y. Wei, C. Szekeres, M. C. Kugler, P. J. Wolters, M. L. Hill, J. A. Frank, A. N. Brumwell, S. E. Wheeler, J. A. Kreidberg, and H. A. Chapman. 2009. Epithelial cell α3 β1 integrin links β-catenin and Smad signaling to promote myofibroblast formation and pulmonary fibrosis. *Journal of Clinical Investigation* 119 (1):213–224. doi: 10.1172/Jci36940.

Kobayashi, H. and M. W. Brechbiel. 2005. Nano-sized MRI contrast agents with dendrimer cores. *Advanced Drug Delivery Reviews* 57 (15):2271–2286. doi: 10.1016/j.addr.2005.09.016.

Kobayashi, T., T. Ishida, Y. Okada, S. Ise, H. Harashima, and H. Kiwada. 2007. Effect of transferrin receptor-targeted liposomal doxorubicin in P-glycoprotein-mediated drug resistant tumor cells. *International Journal of Pharmaceutics* 329 (1–2):94–102. doi: 10.1016/j.ijpharm.2006.08.039.

Kobilka, B. K. 2007. G protein coupled receptor structure and activation. *Biochimica et Biophysica Acta—Biomembranes* 1768 (4):794–807. doi: 10.1016/j.bbamem.2006.10.021.

Kontoyianni, M. and Z. Liu. 2012. Structure-based design in the GPCR target space. *Current Medicinal Chemistry* 19 (4):544–556.

Koshy, V. and P. S. Avasthi. 1987. The anionic sites at luminal surface of peritubular capillaries in rats. *Kidney International* 31 (1):52–58.

Kratz, F. 2008. Albumin as a drug carrier: Design of prodrugs, drug conjugates and nanoparticles. *Journal of Controlled Release* 132 (3):171–183. doi: 10.1016/j.jconrel.2008.05.010.

Krueger, S., T. Hundertmark, D. Kuester, T. Kalinski, U. Peitz, and A. Roessner. 2007. *Helicobacter pylori* alters the distribution of ZO-1 and p120ctn in primary human gastric epithelial cells. *Pathology Research and Practice* 203 (6):433–444. doi: 10.1016/j.prp.2007.04.003.

Laakkonen, J. P., A. R. Makela, E. Kakkonen, P. Turkki, S. Kukkonen, J. Peranen, S. Yla-Herttuala, K. J. Airenne, C. Oker-Blom, M. Vihinen-Ranta, and V. Marjomaki. 2009. Clathrin-independent entry of baculovirus triggers uptake of *E. coli* in non-phagocytic human cells. *PLoS ONE* 4 (4):E5093. doi: 10.1371/Journal.Pone.0005093.

Lagerstrom, M. C. and H. B. Schioth. 2008. Structural diversity of G protein-coupled receptors and significance for drug discovery. *Nature Reviews Drug Discovery* 7 (4):339–357. doi: 10.1038/nrd2518.

Landsiedel, R., E. Fabian, L. Ma-Hock, W. Wohlleben, K. Wiench, F. Oesch, and B. van Ravenzwaay. 2012. Toxico-/biokinetics of nanomaterials. *Archives of Toxicology* 86 (7):1021–1060. doi: 10.1007/s00204-012-0858-7.

Last, J. A., S. J. Liliensiek, P. F. Nealey, and C. J. Murphy. 2009. Determining the mechanical properties of human corneal basement membranes with atomic force microscopy. *Journal of Structural Biology* 167 (1):19–24. doi: 10.1016/j.jsb.2009.03.012.

Lee, H. S., K. Namkoong, D. H. Kim, K. J. Kim, Y. H. Cheong, S. S. Kim, W. B. Lee, and K. Y. Kim. 2004. Hydrogen peroxide-induced alterations of tight junction proteins in bovine brain microvascular endothelial cells. *Microvascular Research* 68 (3):231–238. doi: 10.1016/j.mvr.2004.07.005.

Lefkowitz, R. J. 2007. Seven transmembrane receptors: Something old, something new. *Acta Physiologica* 190 (1):9–19. doi: 10.1111/j.1748-1716.2007.01693.x.

Liliensiek, S. J., P. Nealey, and C. J. Murphy. 2009. Characterization of endothelial basement membrane nanotopography in rhesus macaque as a guide for vessel tissue engineering. *Tissue Engineering Part A* 15 (9):2643–2651. doi: 10.1089/ten.tea.2008.0284.

Liu, X. F., W. L. Jin, and E. C. Theil. 2003. Opening protein pores with chaotropes enhances Fe reduction and chelation of Fe from the ferritin biomineral. *Proceedings of the National Academy of Sciences of the United States of America* 100 (7):3653–3658. doi: 10.1073/pnas.0636928100.

Lu, Z., L. Ding, H. Hong, J. Hoggard, Q. Lu, and Y. H. Chen. 2011. Claudin-7 inhibits human lung cancer cell migration and invasion through ERK/MAPK signaling pathway. *Experimental Cell Research* 317 (13):1935–1946. doi: 10.1016/j.yexcr.2011.05.019.

Lum, S. S., W. S. Fletcher, M. S. O'Dorisio, R. W. Nance, R. F. Pommier, and M. Caprara. 2001. Distribution and functional significance of somatostatin receptors in malignant melanoma. *World Journal of Surgery* 25 (4):407–412.

Maeda, H. and Y. Matsumura. 1989. Tumoritropic and lymphotropic principles of macromolecular drugs. *Critical Reviews in Therapeutic Drug Carrier Systems* 6 (3):193–210.

Maeda, H. and Y. Matsumura. 2011. EPR effect based drug design and clinical outlook for enhanced cancer chemotherapy preface. *Advanced Drug Delivery Reviews* 63 (3):129–130. doi: 10.1016/j.addr.2010.05.001.

Maeda, H., J. Wu, T. Sawa, Y. Matsumura, and K. Hori. 2000. Tumor vascular permeability and the EPR effect in macromolecular therapeutics: A review. *Journal of Controlled Release* 65 (1–2):271–284.

Maitani, Y. 1996. Physico-chemical characterization and application for drug carrier of soybean-derived sterols and their glucosides-containing liposomes. *Yakugaku Zasshi—Journal of the Pharmaceutical Society of Japan* 116 (12):901–910.

Maitani, Y., Y. Nakamura, M. Kon, E. Sanada, K. Sumiyoshi, N. Fujine, M. Asakawa, M. Kogiso, and T. Shimizu. 2013. Higher lung accumulation of intravenously injected organic nanotubes. *International Journal of Nanomedicine* 8:315–23. doi: 10.2147/IJN.S38462.

Martin, T. A., R. E. Mansel, and W. G. Jiang. 2010. Loss of occludin leads to the progression of human breast cancer. *International Journal of Molecular Medicine* 26 (5):721–732. doi: 10.3892/ijmm_00000519.

Matei, D. V., G. Renne, M. Pimentel, M. T. Sandri, L. Zorzino, E. Botteri, C. De Cicco et al. 2012. Neuroendocrine differentiation in castration-resistant prostate cancer: A systematic diagnostic attempt. *Clinical Genitourinary Cancer* 10 (3):164–173. doi: 10.1016/j.clgc.2011.12.004.

Matsumura, Y. 2008. Poly (amino acid) micelle nanocarriers in preclinical and clinical studies. *Advanced Drug Delivery Reviews* 60 (8):899–914. doi: 10.1016/j.addr.2007.11.010.

Matsumura, Y. and H. Maeda. 1986. A new concept for macromolecular therapeutics in cancer-chemotherapy—Mechanism of tumoritropic accumulation of proteins and the antitumor agent smancs. *Cancer Research* 46 (12):6387–6392.

Maul, G. G. 1971. Structure and formation of pores in fenestrated capillaries. *Journal of Ultrastructure Research* 36:768–782.

Maxwell, G. D. and P. D. Sietz. 1985. Development of cells containing catecholamines and somatostatin-like immunoreactivity in neural crest cultures—Relationship of DNA-synthesis to phenotypic-expression. *Developmental Biology* 108 (1): 203–209.

McDonald, D. M. and P. Baluk. 2002. Significance of blood vessel leakiness in cancer. *Cancer Research* 46:6387–6392.

Montes-Burgos, I., D. Walczyk, P. Hole, J. Smith, I. Lynch, and K. Dawson. 2010. Characterisation of nanoparticle size and state prior to nanotoxicological studies. *Journal of Nanoparticle Research* 12 (1):47–53. doi: 10.1007/s11051-009-9774-z.

Morikawa, S., P. Baluk, T. Kaidoh, A. Haskell, R. K. Jain, and D. M. McDonald. 2002. Abnormalities in pericytes on blood vessels and endothelial sprouts in tumors. *The American Journal of Pathology* 160 (3):985–1000. doi: 10.1016/S0002-9440(10)64920-6.

Mudhakir, D. and H. Harashima. 2009. Learning from the viral journey: How to enter cells and how to overcome intracellular barriers to reach the nucleus. *AAPS Journal* 11 (1):65–77. doi: 10.1208/s12248-009-9080-9.

Muto, M. 1976. A scanning and transmission electron microscopic study on rat bone marrow sinuses and transmural migration of blood cells. *Archives of Histology and Cytology* 39:369–386.

Nishiyama, N. and K. Kataoka. 2001. Preparation and characterization of size-controlled polymeric micelle containing cis-dichlorodiammineplatinum(II) in the core. *Journal of Controlled Release* 74 (1–3):83–94.

Nishiyama, N., S. Okazaki, H. Cabral, M. Miyamoto, Y. Kato, Y. Sugiyama, K. Nishio, Y. Matsumura, and K. Kataoka. 2003. Novel cisplatin-incorporated polymeric micelles can eradicate solid tumors in mice. *Cancer Research* 63 (24): 8977–8983.

Nishiyama, N., M. Yokoyama, T. Aoyagi, T. Okano, Y. Sakurai, and K. Kataoka. 1999. Preparation and characterization of self-assembled polymer-metal complex micelle from cis-dichlorodiammineplatinum(II) and poly(ethylene glycol)-poly(α,β-aspartic acid) block copolymer in an aqueous medium. *Langmuir* 15 (2): 377–383.

Obata, S. and K. Honda. 2011. Dynamic behavior of carbon nanotube and bio-/artificial surfactants complexes in an aqueous environment. *Journal of Physical Chemistry C* 115 (40):19659–19667. doi: 10.1021/Jp2072809.

Ohland, C. L. and W. K. MacNaughton. 2010. Probiotic bacteria and intestinal epithelial barrier function. *American Journal of Physiology—Gastrointestinal and Liver Physiology* 298 (6):G807–G819. doi: 10.1152/ajpgi.00243.2009.

Ono, T., R. Tsuruta, M. Fujita, H. S. Aki, S. Kutsuna, Y. Kawamura, J. Wakatsuki et al. 2009. Xanthine oxidase is one of the major sources of superoxide anion radicals in blood after reperfusion in rats with forebrain ischemia/reperfusion. *Brain Research* 1305:158–167. doi: 10.1016/j.brainres.2009.09.061.

Oomen, S. P. M. A., E. G. R. Lichtenauer-Kaligis, N. Verplanke, J. Hofland, S. W. J. Lamberts, B. Lowenberg, and I. P. Touw. 2001. Somatostatin induces migration of acute myeloid leukemia cells via activation of somatostatin receptor subtype 2. *Leukemia* 15 (4):621–627.

Orlando, C., C. C. Raggi, S. Bianchi, V. Distante, L. Simi, V. Vezzosi, S. Gelmini et al. 2004. Measurement of somatostatin receptor subtype 2 mRNA in breast cancer and corresponding normal tissue. *Endocrine-Related Cancer* 11 (2):323–332.

Papotti, M., S. Croce, M. Bello, M. Bongiovanni, E. Allia, M. Schindler, and G. Bussolati. 2001. Expression of somatostatin receptor types 2, 3 and 5 in biopsies and surgical specimens of human lung tumours—Correlation with preoperative octreotide scintigraphy. *Virchows Archiv* 439 (6):787–797. doi: 10.1007/s004280100494.

Passarelli, A. L. 2011. Barriers to success: How baculoviruses establish efficient systemic infections. *Virology* 411 (2):383–392. doi: 10.1016/j.virol.2011.01.009.

Patel, Y. C. 1999. Somatostatin and its receptor family. *Frontiers in Neuroendocrinology* 20 (3):157–198.

Peters, R. 2009. Functionalization of a nanopore: The nuclear pore complex paradigm. *Biochimica et Biophysica Acta—Molecular Cell Research* 1793 (10):1533–1539. doi: 10.1016/j.bbamcr.2009.06.003.

Phan, G., A. Herbet, S. Cholet, H. Benech, J. R. Deverre, and E. Fattal. 2005. Pharmacokinetics of DTPA entrapped in conventional and long-circulating liposomes of different size for plutonium decorporation. *Journal of Controlled Release* 110 (1):177–188. doi: 10.1016/j.jconrel.2005.09.029.

Plummer, R., R. H. Wilson, H. Calvert, A. V. Boddy, M. Griffin, J. Sludden, M. J. Tilby et al. 2011. A phase I clinical study of cisplatin-incorporated polymeric micelles (NC-6004) in patients with solid tumours. *British Journal of Cancer* 104 (4):593–598. doi: 10.1038/Bjc.2011.6.

Prabhune, M., G. Belge, A. Dotzauer, J. Bullerdiek, and M. Radmacher. 2012. Comparison of mechanical properties of normal and malignant thyroid cells. *Micron* 43 (12):1267–1272. doi: 10.1016/j.micron.2012.03.023.

Qian, Z. M., H. Y. Li, H. Z. Sun, and K. Ho. 2002. Targeted drug delivery via the transferrin receptor-mediated endocytosis pathway. *Pharmacological Reviews* 54 (4):561–587.

Rejman, J., V. Oberle, I. S. Zuhorn, and D. Hoekstra. 2004. Size-dependent internalization of particles via the pathways of clathrin-and caveolae-mediated endocytosis. *Biochemical Journal* 377:159–169. doi: 10.1042/Bj20031253.

Reubi, J. C., B. Waser, J. C. Schaer, and J. A. Laissue. 2001. Somatostatin receptor sst1-sst5 expression in normal and neoplastic human tissues using receptor autoradiography with subtype-selective ligands. *European Journal of Nuclear Medicine* 28 (7):836–846.

Rodewald, R. and M. J. Karnovsk. 1974. Porous substructure of glomerular slit diaphragm in rat and mouse. *Journal of Cell Biology* 60 (2):423–433. doi: 10.1083/jcb.60.2.423.

Ross, M. H. and W. Pawlina. 2011. *Histology: A Text and Atlas*, 6th edn. Baltimore, MD: Lippincott Williams & Wilkins.

Sabe, H. 2011. Cancer early dissemination: Cancerous epithelial-mesenchymal transdifferentiation and transforming growth factor β signalling. *Journal of Biochemistry* 149 (6):633–639. doi: 10.1093/Jb/Mvr044.

Scheele, S., A. Nystrom, M. Durbeej, J. F. Talts, M. Ekblom, and P. Ekblom. 2007. Laminin isoforms in development and disease. *Journal of Molecular Medicine (JMM)* 85 (8):825–836. doi: 10.1007/s00109-007-0182-5.

Schmid-Schonbein, G. W. 1990. Mechanisms causing initial lymphatics to expand and compress to promote lymph flow. *Archives of Histology and Cytology* 53 (Suppl):107–114.

Schneeberger, E. E. and R. D. Lynch. 2004. The tight junction: A multifunctional complex. *American Journal of Physiology—Cell Physiology* 286 (6):C1213–C1228. doi: 10.1152/ajpcell.00558.2003.

Scotton, C. J. and R. C. Chambers. 2007. Molecular targets in pulmonary fibrosis—The myofibroblast in focus. *Chest* 132 (4):1311–1321. doi: 10.1378/chest.06-2568.

Sharma, H. S., S. F. Ali, S. M. Hussain, J. J. Schlager, and A. Sharma. 2009. Influence of engineered nanoparticles from metals on the blood-brain barrier permeability, cerebral blood flow, brain edema and neurotoxicity. An experimental study in the rat and mice using biochemical and morphological approaches. *Journal of Nanoscience and Nanotechnology* 9 (8):5055–5072. doi: 10.1166/jnn.2009.GR09.

Sharma, H. S., S. Hussain, J. Schlager, S. F. Ali, and A. Sharma. 2010. Influence of nanoparticles on blood–brain barrier permeability and brain edema formation in rats. *Brain Edema XIV, Acta Neurochirurgica Supplementum* 106:359–364. doi: 10.1007/978-3-211-98811-4_65.

Shimizu, K., Y. Maitani, K. Takayama, and T. Nagai. 1996. Evaluation of dipalmitoylphosphatidylcholine liposomes containing a soybean-derived sterylglucoside mixture for liver targeting. *Journal of Drug Targeting* 4 (4):245–253.

Shiraishi, K., Y. Harada, K. Kawano, Y. Maitani, K. Hori, K. Yanagihara, M. Takigahira, and M. Yokoyama. 2012. Tumor environment changed by combretastatin derivative (Cderiv) pretreatment that leads to effective tumor targeting, MRI studies, and antitumor activity of polymeric micelle carrier systems. *Pharmaceutical Research* 29 (1):178–186. doi: 10.1007/s11095-011-0525-3.

Shiraishi, K., K. Kawano, Y. Maitani, and M. Yokoyama. 2010. Polyion complex micelle MRI contrast agents from poly(ethylene glycol)-b-poly(L-lysine) block copolymers having Gd-DOTA; preparations and their control of T(1)-relaxivities and blood circulation characteristics. *Journal of Controlled Release* 148 (2):160–167. doi: 10.1016/j.jconrel.2010.08.018.

Shiraishi, K., K. Kawano, T. Minowa, Y. Maitani, and M. Yokoyama. 2009. Preparation and in vivo imaging of PEG-poly(L-lysine)-based polymeric micelle MRI contrast agents. *Journal of Controlled Release* 136 (1):14–20. doi: 10.1016/j.jconrel.2009.01.010.

Shirato, I., Y. Tomino, H. Koide, and T. Sakai. 1991. Fine-structure of the glomerular-basement-membrane of the rat-kidney visualized by high-resolution scanning electron-microscopy. *Cell and Tissue Research* 266 (1):1–10.

Shukla, M. N., J. L. Rose, R. Ray, K. L. Lathrop, A. Ray, and P. Ray. 2009. Hepatocyte growth factor inhibits epithelial to myofibroblast transition in lung cells via Smad7. *American Journal of Respiratory Cell and Molecular Biology* 40 (6):643–653. doi: 10.1165/rcmb.2008-0217OC.

Soini, Y. 2011. Claudins in lung diseases. *Respiratory Research* 12:70. doi: 10.1186/1465-9921-12-70.

Spear, M., J. Guo, A. Turner, D. Y. Yu, W. F. Wang, B. Meltzer, S. J. He, X. H. Hu, H. Shang, J. Kuhn, and Y. T. Wu. 2014. HIV-1 triggers WAVE2 phosphorylation in primary CD4 T cells and macrophages, mediating Arp2/3-dependent nuclear migration. *Journal of Biological Chemistry* 289 (10):6949–6959. doi: 10.1074/jbc.M113.492132.

Stan, R. V. 2004. Multiple PV1 dimers reside in the same stomatal or fenestral diaphragm. *American Journal of Physiology—Heart and Circulatory Physiology* 286 (4): H1347–H1353. doi: 10.1152/ajpheart.00909.2003.

Stan, R. V. 2007. Resolved: Capillary endothelium is a major contributor to the glomerular filtration barrier. *Journal of the American Society of Nephrology* 18 (9):2432–2438. doi: 10.1681/Asn.2007060687.

Stan, R. V., E. Tkachenko, and I. R. Niesman. 2004. PV1 is a key structural component for the formation of the stomatal and fenestral diaphragms. *Molecular Biology of the Cell* 15 (8):3615–3630. doi: 10.1091/mbc.E03-08-0593.

Standring, S. 2008. *Gray's Anatomy*, 40th edn. London: Churchill Livingstone, Elsevier.

Suzuki, R., T. Takizawa, Y. Kuwata, M. Mutoh, N. Ishiguro, N. Utoguchi, A. Shinohara, M. Eriguchi, H. Yanagie, and K. Maruyama. 2008. Effective anti-tumor activity of oxaliplatin encapsulated in transferrin-PEG-liposome. *International Journal of Pharmaceutics* 346 (1–2):143–150. doi: 10.1016/j.ijpharm.2007.06.010.

Takeichi, M. 1991. Cadherin cell-adhesion receptors as a morphogenetic regulator. *Science* 251 (5000):1451–1455. doi: 10.1126/science.2006419.

Takeichi, M. 1993. Cadherins in cancer: Implications for invasion and metastasis. *Current Opinion in Cell Biology* 5 (5):806–811. doi: 10.1016/0955-0674(93)90029-P.

Takeichi, M. 1995. Morphogenetic roles of classic cadherins. *Current Opinion in Cell Biology* 7 (5):619–627. doi: 10.1016/0955-0674(95)80102-2.

Tani, E., S. Yamagata, and Y. Ito. 1977. Freeze-fracture of capillary endothelium in rat-brain. *Cell and Tissue Research* 176 (2):157–165.

Tenzer, S., D. Docter, S. Rosfa, A. Wlodarski, J. Kuharev, A. Rekik, S. K. Knauer et al. 2011. Nanoparticle size is a critical physicochemical determinant of the human blood plasma corona: A comprehensive quantitative proteomic analysis. *ACS Nano* 5 (9):7155–7167. doi: 10.1021/Nn201950e.

Thakral, C. and J. L. Abraham. 2009. Gadolinium-induced nephrogenic systemic fibrosis is associated with insoluble Gd deposits in tissues: In vivo transmetallation confirmed by microanalysis. *Journal of Cutaneous Pathology* 36 (12):1244–1254. doi: 10.1111/j.1600-0560.2009.01283.x.

Timpl, R. and J. C. Brown. 1994. The laminins. *Matrix Biology* 14 (4):275–281. doi: 10.1016/0945-053x(94)90192-9.

Tsukita, S. 2001. Claudins and the tight junction barrier. *Kidney International* 60 (2):407.

Tsukita, S., M. Furuse, K. Furuse, and H. Sasaki. 2001. Conversion of Zonulae occludentes from tight to leaky strand type by introducing claudin-2 into Madin-Darby canine kidney I cells. *Journal of Cell Biology* 153 (2):263–272.

Tzu, J. and M. P. Marinkovich. 2008. Bridging structure with function: Structural, regulatory, and developmental role of laminins. *International Journal of Biochemistry & Cell Biology* 40 (2):199–214. doi: 10.1016/j.biocel.2007.07.015.

Uchino, H., Y. Matsumura, T. Negishi, F. Koizumi, T. Hayashi, T. Honda, N. Nishiyama, K. Kataoka, S. Naito, and T. Kakizoe. 2005. Cisplatin-incorporating polymeric micelles (NC-6004) can reduce nephrotoxicity and neurotoxicity of cisplatin in rats. *British Journal of Cancer* 93 (6):678–687. doi: 10.1038/sj.bjc.6602772.

Ulluwishewa, D., R. C. Anderson, W. C. McNabb, P. J. Moughan, J. M. Wells, and N. C. Roy. 2011. Regulation of tight junction permeability by intestinal bacteria and dietary components. *Journal of Nutrition* 141 (5):769–776. doi: 10.3945/jn.110.135657.

van der Goes, A., D. Wouters, S. M. A. van der Pol, R. Huizinga, E. Ronken, P. Adamson, J. Greenwood, C. D. Dijkstra, and H. E. de Vries. 2001. Reactive oxygen species enhance the migration of monocytes across the blood-brain barrier in vitro. *FASEB Journal* 15 (8):1852–1862. doi: 10.1096/fj.00-0881fje.

Van Itallie, C. M. and J. M. Anderson. 2006. Claudins and epithelial paracellular transport. *Annual Review of Physiology* (Palo Alto) 68:403–429.

Van Itallie, C. M., J. Holmes, A. Bridges, J. L. Gookin, M. R. Coccaro, W. Proctor, O. R. Colegio, and J. M. Anderson. 2008. The density of small tight junction pores varies among cell types and is increased by expression of claudin-2. *Journal of Cell Science* 121 (3):298–305. doi: 10.1242/Jcs.021485.

Wang, J., U. B. Jensen, G. V. Jensen, S. Shipovskov, V. S. Balakrishnan, D. Otzen, J. S. Pedersen, F. Besenbacher, and D. S. Sutherland. 2011. Soft interactions at nanoparticles alter protein function and conformation in a size dependent manner. *Nano Letters* 11 (11):4985–4991. doi: 10.1021/Nl202940k.

Wartiovaara, J., L. G. Ofverstedt, J. Khoshnoodi, J. J. Zhang, E. Makela, S. Sandin, V. Ruotsalainen, R. H. Cheng, H. Jalanko, U. Skoglund, and K. Tryggvason. 2004. Nephrin strands contribute to a porous slit diaphragm scaffold as revealed by electron tomography. *Journal of Clinical Investigation* 114 (10):1475–1483. doi: 10.1172/jci200422562.

Wayengera, M. 2010. On the general theory of the origins of retroviruses. *Theoretical Biology and Medical Modelling* 7:5. doi: 10.1186/1742-4682-7-5.

Weinberg, R. A. 2007. *The Biology of Cancer.* New York: Garland Science, Taylor & Francis Group.

Willis, B. C., J. M. Liebler, K. Luby-Phelps, A. G. Nicholson, E. D. Crandall, R. M. du Bois, and Z. Borok. 2005. Induction of epithelial-mesenchymal transition in alveolar epithelial cells by transforming growth factor-ss 1—Potential role in idiopathic pulmonary fibrosis. *American Journal of Pathology* 166 (5):1321–1332. doi: 10.1016/S0002-9440(10)62351-6.

Xia, X. R., N. A. Monteiro-Riviere, S. Mathur, X. F. Song, L. S. Xiao, S. J. Oldenberg, B. Fadeel, and J. E. Riviere. 2011. Mapping the surface adsorption forces of nanomaterials in biological systems. *ACS Nano* 5 (11):9074–9081. doi: 10.1021/Nn203303c.

Xia, X. R., N. A. Monteiro-Riviere, and J. E. Riviere. 2010. An index for characterization of nanomaterials in biological systems. *Nature Nanotechnology* 5 (9):671–675. doi: 10.1038/Nnano.2010.164.

Yamada, Y. 1994. *Modern Textbook of Histology, Third Edition*. Tokyo: Kanehara & Co., Ltd.

Yamato, M., T. Egashira, and H. Utsumi. 2003. Application of in vivo ESR spectroscopy to measurement of cerebrovascular ROS generation in stroke. *Free Radical Biology and Medicine* 35 (12):1619–1631. doi: 10.1016/j.freeradbiomed. 2003.09.013.

Yamazaki, Y., R. Tokumasu, H. Kimura, and S. Tsukita. 2011. Role of claudin species-specific dynamics in reconstitution and remodeling of the zonula occludens. *Molecular Biology of the Cell* 22 (9):1495–1504. doi: 10.1091/mbc.E10-12-1003.

Yasuhara, N., M. Oka, and Y. Yoneda. 2009. The role of the nuclear transport system in cell differentiation. *Seminars in Cell & Developmental Biology* 20 (5):590–599. doi: 10.1016/j.semcdb.2009.05.003.

Yokoyama, M., M. Miyauchi, N. Yamada, T. Okano, Y. Sakurai, K. Kataoka, and S. Inoue. 1990. Polymer micelles as novel drug carrier—Adriamycin-conjugated poly(ethylene glycol) poly(aspartic acid) block copolymer. *Journal of Controlled Release* 11 (1–3):269–278.

Yokoyama, M., T. Okano, Y. Sakurai, H. Ekimoto, C. Shibazaki, and K. Kataoka. 1991. Toxicity and antitumor-activity against solid tumors of micelle-forming polymeric anticancer drug and its extremely long circulation in blood. *Cancer Research* 51 (12):3229–3236.

Yokoyama, M., T. Okano, Y. Sakurai, S. Suwa, and K. Kataoka. 1996. Introduction of cisplatin into polymeric micelle. *Journal of Controlled Release* 39 (2–3):351–356.

Zhang, H. Z., K. E. Burnum, M. L. Luna, B. O. Petritis, J. S. Kim, W. J. Qian, R. J. Moore et al. 2011. Quantitative proteomics analysis of adsorbed plasma proteins classifies nanoparticles with different surface properties and size. *Proteomics* 11 (23):4569–4577. doi: 10.1002/pmic.201100037.

Zhang, L. W. and N. A. Monteiro-Riviere. 2009. Mechanisms of quantum dot nanoparticle cellular uptake. *Toxicological Sciences* 110 (1):138–155. doi: 10.1093/toxsci/kfp087.

Zhang, Y. A., X. Q. Wang, J. C. Wang, X. A. Zhang, and Q. A. Zhang. 2011. Octreotide-modified polymeric micelles as potential carriers for targeted docetaxel delivery to somatostatin receptor overexpressing tumor cells. *Pharmaceutical Research* 28 (5):1167–1178. doi: 10.1007/s11095-011-0381-1.

Zolnik, B. S. and N. Sadrieh. 2009. Regulatory perspective on the importance of ADME assessment of nanoscale material containing drugs. *Advanced Drug Delivery Reviews* 61 (6):422–427. doi: 10.1016/j.addr.2009.03.006.

Zolnik, B. S., S. T. Stern, J. M. Kaiser, Y. Heakal, J. D. Clogston, M. Kester, and S. E. McNeil. 2008. Rapid distribution of liposomal short-chain ceramide in vitro and in vivo. *Drug Metabolism and Disposition* 36 (8):1709–1715. doi: 10.1124/dmd.107.019679.

Zyrek, A. A., C. Cichon, S. Helms, C. Enders, U. Sonnenborn, and M. A. Schmidt. 2007. Molecular mechanisms underlying the probiotic effects of *Escherichia coli* Nissle 1917 involve ZO-2 and PKC zeta redistribution resulting in tight junction and epithelial barrier repair. *Cellular Microbiology* 9 (3):804–816. doi: 10.1111/j.1462-5822.2006.00836.x.

7

Toxicology in the Nanoscale World

The toxicology of nanoproducts in the nanoscale world differs from that of conventional substances in terms of basic toxicological principles and, indeed, may require new concepts of toxicological principles and testing. In this chapter, the unique characteristics of nanoproducts are discussed, including exposure to toxic nanoscale substances, the spectrum of undesired effects, tolerance, dose response, and descriptive animal toxicity testing.

7.1 Characteristics of Exposure

Low-molecular-weight (LMW) conventional drugs or materials can easily cross the permeability barriers set up by the body. However, nanoscale materials cannot readily travel through the paracellular pathway afforded by tight junctions and cell–cell adhesions. Even if nanomaterials enter the body, their distribution and accumulation are restricted. The sole exception is nanoparticle penetration into local tissue damaged by injury or disease. Under normal conditions, the paracellular pathway only allows free access to nanoproducts when the permeability barrier is temporarily breached by, for example, excessive amounts of free radicals or reactive oxygen species.

The concentration of active ingredients in the body is equivalent to exposure for conventional materials, while the concentration of active ingredients together with intact particles and platform affects exposure to nanomaterials. In the case of biodegradable passive-targeting nanomedicines, both intact nanoparticles and degraded fragments can be endocytosed into the target cell. Therefore, true exposure includes released active ingredients, intact nanoparticles, and degraded fragments. By this reasoning, a conventional drug or substance enables quantitative interpretation of exposure by the concentration of active ingredients, whereas nanoproducts modify the quantitative interpretation of exposure due to biodegradation over time. Therefore, cell exposure to conventional substances can be estimated by the dose, while cell exposure to active-targeting nanomedicines or nonbiodegradable nanoparticles cannot, because both the concentration of the nanoparticle and the quantity of active ingredients per nanoparticle become factored into the dose.

Conventional drugs contain one active ingredient per molecule, permitting homogeneity in dispersion before administration to the body, as well as homogeneity in the body after administration. However, nanomedicines are polydisperse materials with varying quantities of active ingredients per particle, thereby restricting homogeneity. Because the distance between molecules in conventional drugs is small, homogeneity is easily maintained, even with repeated dilution (Figure 7.1a). By contrast, variations in dispersion are common with repeated dilution of nanomedicines (Figure 7.1a). Furthermore, nanomedicines are generally coated with polyethylene glycol (PEG) or ligands. The coating of the nanoparticle depends on the surface area or dimensions of the inner core, and the spaces between

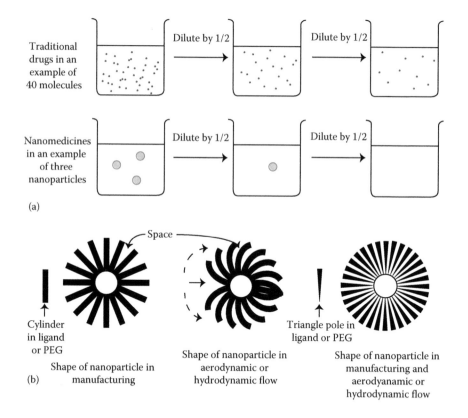

FIGURE 7.1

Characteristics of nanomedicines affecting exposure. (a) Dilution of conventional products and nanoproducts and (b) shape of nanoparticles in manufacturing, as affected by aerodynamic and hydrodynamic flow processes in the body. The diluted concentration of conventional drugs is easily maintained by the homogeneity in an example of twice half dilution. However, the diluted concentration of nanomedicines may be not often maintained by the dispersion in solution as well as interaction the coating of the nanoparticle and the surface of coarse substrates of experimental instruments even in an example of twice half dilution.

PEG and ligand molecules change with aerodynamic and hydraulic flow. PEG and ligand spacing is also affected by whether the molecules assume a cylindrical or a triangular shape (Figure 7.1b). Theoretically, nanoparticles easily adsorb not only onto cells by surface interactions and aerodynamic or hydrodynamic flow but also onto coarse substrates, such as those provided by experimental instruments.

Nanoparticle size is analyzed by dynamic light scattering (DLS) measurement during manufacture. The nominal size in DLS analysis is based on the Brownian motion under the assumption of a spherical shape. Therefore, nanoparticles with an elliptical shape are analyzed by the Brownian motion according to neither the longest nor the shortest diameter in the aspect ratio.

7.1.1 Route and Site of Exposure

The main routes of exposure of foreign substances into the body include oral administration into the buccal cavity and the GI tract, as well as intravenous injection into blood vessels, inhalation into the respiratory and olfactory systems, administration into the integumentary system via the pilosebaceous unit or the skin, and intraocular administration into the eye. Although conventional drugs and nanoproducts utilize the same general routes, the site of exposure can differ at the level of cellular access. Conventional drugs are mainly taken up by intestinal absorptive cells in the GI tract, while nanoproducts are mainly taken up in intact form by intestinal M cells and neuroendocrine cells.

7.1.2 Dose and Loading Rate

The dose of a conventional product is determined by the weight or quantity of the active ingredients. The weight of active ingredients per unit body weight is expressed as the dose. The dose of a nanoproduct is determined by the weight or quantity of active ingredients per unit body weight for passive-targeting nanomedicines and by the weight of active ingredients plus the weight of ligand per unit body weight for active-targeting nanomedicines. For nanoproduct dose, we must also clarify the loading rate, as described in the following.

One nanoparticle is the sum of the platform and the active ingredients. The loading rate (%) signifies product efficiency and is used for the calculation of the manufacturing cost (see the formula for loading rate). The loading rate (%) as the content of active ingredients is calculated for both the dose by active ingredients and the dose by nanoparticle weight. The weight of the nanoparticle is also essential for calculating the maximal feasible dose (MFD, generally 1–2 g/kg), which is further discussed in Section 7.2.1 (allergic reactions). The loading rate (%) (as ligand content for the dose by ligand) is calculated if the ligand has pharmacological action.

$$\text{Loading rate as manufacturing} = \frac{\text{Nanoparticles (g)}}{\text{Used substances (g)}}$$

Loading rate as content of active ingredients

$$= \frac{\text{Active ingredients (g) per one nanoparticle}}{\text{One nanoparticle (g)}}$$

$$\text{Loading rate as content of ligand} = \frac{\text{Ligands (g) per one nanoparticle}}{\text{One nanoparticle (g)}}$$

7.1.3 Duration and Frequency of Exposure

Conventional acute exposure is defined as exposure to a chemical compound for <24 h. Subacute, subchronic, and chronic exposures refer to repeated exposure to a chemical compound for ≤1, 1–3, and ≥3 months, respectively. Nanoproducts are not treated the same as conventional drugs, regardless of a common or different drug regimen, due to the elongated duration of nanoproduct exposure. Although administration of biodegraded or active-targeting nanomedicines by intravenous injection constitutes single-dose acute exposure in terms of timing, the duration of exposure is actually single-dose subacute in terms of the release of free drug. Administration of nonbio-degraded nanomaterials (e.g., carbon nanotubes) also constitutes single-dose acute exposure in terms of timing, but the duration of exposure is equivalent to single-dose subchronic exposure.

7.1.4 Control Groups in Nanoproduct Analysis

Control groups for conventional substances comprise the vehicle and the positive control group. The vehicle control for nanoproducts is the platform (polymer or liposome) without active ingredients, while the positive control is the unencapsulated (free) active ingredient. The platform alone should theoretically have the same physiochemical properties as the platform loaded with active ingredients. However, it is often difficult to manufacture the same platform for loaded and unloaded nanoproducts, as in the case of conjugated nanomedicines. Nonetheless, the absence of an appropriate vehicle control does not negate the significance of nanoproduct analysis because the effect of a nanoproduct typically depends on engineering properties rather than platform and especially the breakdown of biodegradable nanomedicines.

Pharmacological, pharmacokinetic, and toxicological comparisons between encapsulated and unencapsulated active ingredients may be useful for advertisement purposes and for the establishment of the starting dose in early clinical studies. However, the significance of such comparisons is ambiguous when it comes to assessing the toxicological profile of a nanomedicine.

The characteristics of exposure to unencapsulated ingredients are different from exposure to encapsulated ingredients in a nanoproduct formulation, and thus the toxicological comparison with conventional drugs is restricted. Furthermore, the acute toxicity of a conventional substance stems from effects induced by the peak drug concentration, or the C_{max}, while the toxicity of a slow-release, biodegradable nanomedicine stems from effects induced by the area under the curve (AUC) and depends on the releasability of free drug. Thus, while the pharmacokinetic profile of a conventional drug after infusion may be similar to that of a slowly released drug, the infusion of the former can put the body under undue stress.

7.2 Spectrum of Undesired Effects

7.2.1 Allergic Reactions

Given that injectable nanomedicines are engineered to include stealth as an essential function, these substances are not typically associated with specific allergic reactions in clinical practice. However, mild allergic reactions often occur even in the case of stealthy nanomaterials because of the quantity or type of platform components. Allergic reactions can often be avoided by changing the regimen from bolus injection to short-term infusion.

Allergic reactions to platform components are induced in a dose-related fashion, similar to allergic reactions resulting from the consumption of non-toxic food components (Perrier and Corthesy, 2011). Conventional LMW drugs contain active ingredients and excipients, while injectable nanomedicines contain platform components, encapsulated active ingredients, and excipients. If the encapsulation rate of the active ingredient is 1%, then the clinical dose of 100 mg/m² as active ingredient corresponds to 10 g/m² of the intact nanomedicine. Thus, the clinical dose easily exceeds the upper limit of the MFD in certain dose regimens.

The MFD is rarely considered for conventional drugs, but the total quantity of platform plus active ingredients must be taken into account when setting the dose regimen or dosage of nanomedicines. Phase I clinical studies have recently been conducted for two biodegradable nanomedicines to evaluate the effect of a short infusion time (60 min) on the required clinical dose, with the goal of lowering the clinical dose to a value below the MFD (Hamaguchi et al., 2007; Plummer et al., 2011).

7.2.2 Immediate versus Delayed Toxicity

For conventional drugs, immediate toxic effects are defined as those that occur or develop rapidly after a single administration, whereas delayed toxic effects are those that appear after a substantial lapse of time. Although toxic effects do not always occur rapidly after a single administration of a nanoproduct, the appearance of delayed adverse events is not necessarily defined as delayed toxicity. True immediate or delayed toxicity can only be addressed by the results of studies using the active ingredients or platform components of a biodegradable nanomedicine.

7.3 Tolerance

Two major mechanisms are responsible for the development of tolerance to drugs or toxicants: one involving a decreased amount of substance reaching the site where the toxic effect is produced (dispositional tolerance) and the other involving a reduced responsiveness to the substance over time (physiological tolerance) (Klaassen, 1995). Nanomedicines for injection are strategically designed to reduce distribution into normal tissue and to increase accumulation into target tissues by the EPR effect. These products have a reverse mechanism of action compared with dispositional tolerance, and this reverse mechanism contributes to the attenuation of multidrug resistance in cancer cells in the passive-targeting nanomedicine platform (Farrell et al., 2011; Shapira et al., 2011; Parhi et al., 2012). For active-targeting nanomedicines, multi-drug resistance is disabled by active ingredients that influence the ABC transporter and associated intracellular signaling pathways in cancer cells (Alison et al., 2012; Fukuda and Schuetz, 2012; Shukla et al., 2012).

7.4 Dose–Response Relationships

The dose–response relationship for a nanoproduct depends on its biodegradability. Biodegradable nanomedicines for injection are pharmacokinetically characterized by two dose–response curves. One dose–response curve corresponds to the concentration of active ingredient released from the nanoparticle and is identical to that for the conventional free drug. The other dose–response curve corresponds to the effects of endocytosed intact nanomedicines and partially degraded fragments. Because intact particles cannot pass through the entire body, their distribution is restricted to the sinusoidal space of the liver, the bone marrow, and the lymphatic system. On the other hand, partially degraded fragments distribute in much the same manner as conventional drugs, depending on their size, and should therefore elicit a systemic reaction. The dose–response relationship for an active-targeting nanomedicine is affected by endocytosis and depends on the type of ligand and the affinity for the target cell.

The therapeutic index of a drug is calculated by the LD_{50}/ED_{50} (median lethal dose/median effective dose). The clinical dosage employs infusion rather than bolus injection to avoid adverse events, even in the case of conventional drugs, if the bolus administration is linked to toxicity. The lethal dose of a conventional drug depends on the C_{max}, while the lethal dose of a nanomedicine avoids acute characteristics because of the slow release of active ingredients (for biodegradable nanomedicines) and the selective interaction with target cells (for active-targeting nanomedicines).

7.5 Descriptive Animal Toxicity Tests

7.5.1 Acute Lethality

Dose selection is an important initial element in acute toxicity nonclinical studies. As described previously, nanomedicines are composed of the platform and active ingredients. The platform affects the MFD according to the weight of its constituents when the percentage of active ingredients in the intact nanomedicine is low. For example, if the fractional weight of active ingredient per nanoparticle is 5%, a dose of 50 mg/kg of active ingredient will correspond to 1 g/kg of intact nanoparticle, or it will be the same as the MFD.

PV-1 is critical for the permeability of fenestrated endothelia, as discussed in Chapter 6 (Stan et al., 1999, 2004; Hnasko et al., 2002; Ichimura et al., 2008; Herrnberger et al., 2012; Tkachenko et al., 2012). Excessive doses of positively charged nutrients such as arginine are lethal to mammalian animals. The lethality stems from the occlusion of capillary pores following PV-1 binding with cationic substances, leading to acute renal failure (Schramm et al., 1994; Tome et al., 1999). The use of cationic nanoparticles in future biochemical investigations should enable further dissection of the toxicological mechanism of PV-1 action.

7.5.2 Repeated Dose Study

Although the dose regimen of conventional substances includes almost continuous administration or daily dosing, this dose regimen is not always ideal for nanomedicines. Toxicity studies of nanomedicines indicate that an intermittent dose regimen might be preferable. Administration of a slow-release biodegradable nanomedicine at bi- or triweekly intervals or every second day may be selected over administration every 24 h because most active ingredients disappear within 48 h. Alternatively, a nonbiodegradable nanomaterial with active ingredients that disappear within 1 month might be administered only once per month.

7.5.3 Genotoxicity

The genotoxicity of nanoproducts depends on the properties of the active ingredients. For biodegradable nanomedicines, the slow release of active ingredients requires longer incubation times with cells than free drugs for in vitro studies of genotoxicity.

On a practical note, formulations requiring dispersion in solution must be carefully prepared and treated for all nanomaterial products and applications to avoid the disappearance of the nanoparticles by dilution or the adsorption of the nanoparticles onto the walls of experimental tubes and manufactured packaging; otherwise, the results of genotoxicity and other tests will be skewed.

7.5.4 Developmental and Reproductive Toxicity

The ability of a nanoparticle to pass through the fenestrae of hepatic capillaries affects systemic circulation, distribution, and toxicity (Igarashi, 2008). Anatomical findings of endothelial fenestrae in fetal and infant rats change over the course of development (Bankston and Pino, 1980; Barbera-Guillem et al., 1986; McCuskey et al., 2003). Bankston and Pino (1980) studied the structural development of the fetal liver sinusoid in rats from gestational days 10–22. Fenestrae with diaphragms were observed in the fetus before gestational day 17, and typical open fenestrae in the adult liver appeared around gestational day 17.

McCuskey et al. (2003) demonstrated the onset of hepatic microvascular heterogeneity after birth and explored its temporal relationship to the development of parenchymal cell plates on postnatal days 4–30 in the rat. Sinusoidal endothelial fenestration was sparse on postnatal day 4, but phagocytic Kupffer cell function was already present. Sinusoidal endothelial fenestrae showed the initial signs of organization as sieve plates at this time. The sieve plates were more abundant by postnatal day 30. Some fenestrae were apparently bridged by diaphragms in 4-day-old pups, while the opening of endothelial fenestrae lacking diaphragms progressively increased with age. Kupffer cells were either interdigitated into the sinusoidal wall or attached to the lumenal surface. Between postnatal days 7 and 14, the Kupffer cells were enlarged and appeared to be activated. Microvilli on the hepatocyte surface also increased in size and number with age, with a dramatic increase between the postnatal days 14 and 30.

Barbera-Guillem et al. (1986) conducted a study of endothelial fenestrae in the developing rat liver on embryonic days 18 and 21, on postnatal days 1 and 5, and in the adult (Barbera-Guillem et al., 1986). The central veins of the fetal and newborn livers had a sinusoidal appearance, showing large fenestrae with sieve plates and gaps of various sizes, ranging from 50 to 3000 nm in diameter. The number of fenestrae of 100 nm or more decreased with age and comprised ≤40% of all fenestrae in the fetus versus ≥60% in the adult. Therefore, although nanoproducts cannot travel through fetal tissues during organogenesis, they can pass into systemic circulation via fetal endothelial fenestrae after organogenesis at a higher rate than in the adult and likewise via endothelial fenestrae in the newborn.

Developmental toxicity is induced by exposure of the embryo to drugs or other substances at a critical phase during organogenesis for structural differentiation. The timing of conventional or nanoproduct administration is therefore a definitive factor for the induction of developmental toxicity. Moreover, drug effects on developmental toxicity increase with elevated exposure, and the sensitivity period lengthens (Schardein, 1976).

For descriptive developmental toxicity, the sensitivity period of the induction of abnormalities or malformations has been reported for various organs, including cardiovascular organs (Okamoto, 1968; Okamoto et al., 1984; Lee and Satow, 1987; Lee et al., 1987; Satow et al., 1987), the brain (Holson et al., 1997; Kuwagata et al., 2007), the palate (Ema et al., 1997; Ikemi et al., 2001; Rogers et al., 2004; Tiboni and Giampietro, 2005), and the axial skeleton (Igarashi, 1998; Menegola et al., 1998). The AUC- and high C_{max}-dependent induction of developmental toxicity by drugs has also been reviewed (Nau, 1986; Sullivan, 1999).

In many cases, maternal toxicity occurs at lower drug doses than fetal toxicity. In general, if maternal toxicity were protected by intoxication of toxic substance via metabolism, fetal toxicity could be reduced by the intoxication. However, the importance of maternal acceptable daily intake levels on teratogenic responses depends on the mechanism of teratogenesis and the presence of key metabolic enzymes. In humans, the activity of hepatic microsomal cytochrome P-450 monooxygenase, an enzyme commonly involved in drug metabolism, is lower than that in mice, rats, and hamsters (Lorenz et al., 1978; Walker, 1978).

Nau (1986) studied the teratogenic effects of valproate (i.e., induction of exencephaly) in mice by using either a single daily injection (to yield high C_{max} values) or a slow infusion via an implanted minipump (to yield high AUC values) (Nau, 1986). While a certain peak plasma level was required for the induction of exencephaly, the AUC value was far less important. Studies on caffeine showed a similar phenomenon in that a critical peak plasma level rather than a particular AUC value was required to engender malformations (Sullivan, 1999). On the other hand, cyclophosphamide- and retinoid-induced teratogenicity was more dependent on the AUC than on the C_{max} (Sullivan, 1999).

The pharmacokinetic mechanism of biodegradable nanoproducts must be considered in terms of the intact nanoparticle and the released active ingredients. Pharmacokinetically, nanoproducts prolong the time of reduced peak drug levels and high AUC values for the unencapsulated active ingredients.

These actions are due to the pharmaceutical design of slow-release nano-medicines rather than the activity of hepatic enzymes.

Intact biodegradable nanoproducts show restricted capacity to cross the placental capillary because the placental pores limit the passage of HMW substances of ≥1000 Da (Mirkin, 1973). The elevation of free radicals or reactive oxygen species might sufficiently open the tight junctions to allow passage of intact biodegradable and nonbiodegradable nanoproducts across the placenta, but there is currently no clear evidence to support the augmentation of these compounds during the short period of organogenesis.

7.6 Nanoproduct Compatibility with the Immune System

Nanoproduct compatibility with the immune system is typically tested for the platform rather than for the active ingredients. Compatibility testing involves an assessment of the interactions between the nanoparticle (or more accurately the platform) and the plasma proteins/blood components, in addition to the propensity of the nanoparticle to provoke hemolysis, thrombus formation, and complement activation (Dobrovolskaia et al., 2008). The scope of platform components includes PEG, linker molecules and ligand in the outer shell, and lipids or polymer in the inner shell. Retained impurities must also be considered for block-polymer micelles and liposome nano-medicines. Common pitfalls are sterility and the presence of endotoxins during synthesis and residual components in the platform (Crist et al., 2013). Nanoparticle interference has also been reported for in vitro limulus amoebocyte lysate assays to assess endotoxin levels (Dobrovolskaia et al., 2009).

References

Alison, M. R., W. R. Lin, S. M. L. Lim, and L. J. Nicholson. 2012. Cancer stem cells: In the line of fire. *Cancer Treatment Reviews* 38 (6):589–598. doi: 10.1016/j.ctrv.2012.03.003.

Bankston, P. W. and R. M. Pino. 1980. The development of the sinusoids of fetal rat-liver—Morphology of endothelial-cells, Kupffer cells, and the transmural migration of blood-cells into the sinusoids. *American Journal of Anatomy* 159 (1):1–15. doi: 10.1002/aja.1001590102.

Barbera-Guillem, E., J. M. Arrue, J. Ballesteros, and F. Vidal-Vanaclocha. 1986. Structural changes in endothelial cells of developing rat liver in the transition from fetal to postnatal life. *Journal of Ultrastructure and Molecular Structure Research* 97:197–206.

Crist, R. M., J. H. Grossman, A. K. Patri, S. T. Stern, M. A. Dobrovolskaia, P. P. Adiseshaiah, J. D. Clogston, and S. E. McNeil. 2013. Common pitfalls in nanotechnology: Lessons learned from NCI's Nanotechnology Characterization Laboratory. *Integrative Biology* 5 (1):66–73. doi: 10.1039/C2ib20117h.

Dobrovolskaia, M. A., P. Aggarwal, J. B. Hall, and S. E. McNeil. 2008. Preclinical studies to understand nanoparticle interaction with the immune system and its potential effects on nanoparticle biodistribution. *Molecular Pharmaceutics* 5 (4):487–495. doi: 10.1021/Mp800032f.

Dobrovolskaia, M. A., A. K. Patri, J. W. Zheng, J. D. Clogston, N. Ayub, P. Aggarwal, B. W. Neun, J. B. Hall, and S. E. McNeil. 2009. Interaction of colloidal gold nanoparticles with human blood: Effects on particle size and analysis of plasma protein binding profiles. *Nanomedicine—Nanotechnology Biology and Medicine* 5 (2):106–117. doi: 10.1016/j.nano.2008.08.001.

Ema, M., A. Harazono, E. Miyawaki, and Y. Ogawa. 1997. Effect of the day of administration on the developmental toxicity of tributyltin chloride in rats. *Archives of Environmental Contamination and Toxicology* 33 (1):90–96.

Farrell, D., K. Ptak, N. J. Panaro, and P. Grodzinski. 2011. Nanotechnology-based cancer therapeutics-promise and challenge-lessons learned through the NCI alliance for nanotechnology in cancer. *Pharmaceutical Research* 28 (2):273–278. doi: 10.1007/s11095-010-0214-7.

Fukuda, Y. and J. D. Schuetz. 2012. ABC transporters and their role in nucleoside and nucleotide drug resistance. *Biochemical Pharmacology* 83 (8):1073–1083. doi: 10.1016/j.bcp.2011.12.042.

Hamaguchi, T., K. Kato, H. Yasui, C. Morizane, M. Ikeda, H. Ueno, K. Muro et al. 2007. A phase I and pharmacokinetic study of NK105, a paclitaxel-incorporating micellar nanoparticle formulation. *British Journal of Cancer* 97 (2):170–176. doi: 10.1038/sj.bjc.6603855.

Herrnberger, L., K. Ebner, B. Junglas, and E. R. Tamm. 2012. The role of plasmalemma vesicle-associated protein (PLVAP) in endothelial cells of Schlemm's canal and ocular capillaries. *Experimental Eye Research* 105:27–33. doi: 10.1016/j.exer.2012.09.011.

Hnasko, R., M. McFarland, and N. Ben-Jonathan. 2002. Distribution and characterization of plasmalemma vesicle protein-1 in rat endocrine glands. *Journal of Endocrinology* 175 (3):649–661.

Holson, R. R., R. A. Gazzara, S. A. Ferguson, S. F. Ali, J. B. Laborde, and J. Adams. 1997. Gestational retinoic acid exposure: A sensitive period for effects on neonatal mortality and cerebellar development. *Neurotoxicology and Teratology* 19 (5):335–346. doi: 10.1016/S0892-0362(97)00039-1.

Ichimura, K., R. V. Stan, H. Kurihara, and T. Sakai. 2008. Glomerular endothelial cells form diaphragms during development and pathologic conditions. *Journal of the American Society of Nephrology* 19 (8):1463–1471. doi: 10.1681/Asn.2007101138.

Igarashi, E. 1998. Anomalies of cartilaginous and ossified axial skeleton in rat fetuses treated with cyclophosphamide: Type, frequency, and stage specificity. *Congenital Anomalies* 38 (1):39–55.

Igarashi, E. 2008. Factors affecting toxicity and efficacy of polymeric nanomedicines. *Toxicology and Applied Pharmacology* 229 (1):121–134. doi: 10.1016/j.taap.2008.02.007.

Ikemi, N., Y. Otani, T. Ikegami, and M. Yasuda. 2001. Palatal ruga anomaly induced by all-trans-retinoic acid in the Crj:SD rat: Possible warning sign of teratogenicity. *Reproductive Toxicology* 15 (1):87–93.

Klaassen, C. D. 1995. *Casarett and Doull's Toxicology: The Basic Science of Poisons*, 5th edn. New York: McGraw-Hill.

Kuwagata, M., T. Ogawa, T. Nagata, and S. Shioda. 2007. The evaluation of early embryonic neurogenesis after exposure to the genotoxic agent 5-bromo-2'-deoxyuridine in mice. *Neurotoxicology* 28 (4):780–789. doi: 10.1016/j.neuro. 2006.07.017.

Lee, J.Y., H. Okuda, S. Tutimoto, and Y. Satow. 1987. Teratogenic effects of ^{60}Co γ rays irradiation on rat embryos. *Annual Report of Research Institute for Radiation Biology and Medicine* 28:167–181.

Lee, J.Y. and Y. Satow. 1987. The teratogenic effects of tritiated water and tritium simulator on rat embryos. *Annual Report of Research Institute for Radiation Biology and Medicine* 28:155–166.

Lorenz, J., H. R. Glatt, R. Fleischmann, R. Ferlinz, and F. Oesch. 1984. Drug-metabolism in man and its relationship to that in 3 rodent species—monooxygenase, epoxide hydrolase, and glutathione S-transferase activities in subcellular-fractions of lung and liver. *Biochemical Medicine* 32 (1):43–56. doi: 10.1016/0006-2944(84)90007-3.

McCuskey, R. S., W. Ekataksin, A. V. LeBouton, J. Nishida, M. K. McCuskey, D. McDonnell, C. Williams, N. W. Bethea, B. Dvorak, and O. Koldovsky. 2003. Hepatic microvascular development in relation to the morphogenesis of hepatocellular plates in neonatal rats. *Anatomical Record Part A—Discoveries in Molecular Cellular and Evolutionary Biology* 275A (1):1019–1030. doi: 10.1002/Ar.A.10117.

Menegola, E., M. L. Broccia, M. Prati, and E. Giavini. 1998. Stage-dependent skeletal malformations induced by valproic acid in rat. *International Journal of Developmental Biology* 42 (1):99–102.

Mirkin, B. L. 1973. Maternal and fetal distribution of drugs in pregnancy. *Clinical Pharmacology & Therapeutics* 14 (4):643–647.

Nau, H. 1986. Transfer of valproic acid and its main active unsaturated metabolite to the gestational tissue: Correlation with neural tube defect formation in the mouse. *Teratology* 33 (1):21–27. doi: 10.1002/tera.1420330105.

Okamoto, N. 1968. Morphology and classification of the cardiovascular anomalies induced by 14.1 MeV neutron irradiation. *Annual Report of Research Institute for Radiation Biology and Medicine* 9:25–42.

Okamoto, N., Y. Satow, J. Y. Lee, H. Sumida, K. Hayakawa, S. Ohdo, and T. Okishima. 1984. Morphology and pathogenesis of the cardiovascular anomalies induced by bis-(dichloroacetyl) diamine in rats. In: J. J. Nora and A. Takao, eds. *Congenital Heart Disease: Causes and Processes*. Mount Kisco, New York: Futura Publishing Co., pp. 199–221.

Parhi, P., C. Mohanty, and S. K. Sahoo. 2012. Nanotechnology-based combinational drug delivery: An emerging approach for cancer therapy. *Drug Discovery Today* 17 (17–18):1044–1052. doi: 10.1016/j.drudis.2012.05.010.

Perrier, C. and B. Corthesy. 2011. Gut permeability and food allergies. *Clinical and Experimental Allergy* 41 (1):20–28. doi: 10.1111/j.1365-2222.2010.03639.x.

Plummer, R., R. H. Wilson, H. Calvert, A. V. Boddy, M. Griffin, J. Sludden, M. J. Tilby et al. 2011. A phase I clinical study of cisplatin-incorporated polymeric micelles (NC-6004) in patients with solid tumours. *British Journal of Cancer* 104 (4): 593–598. doi: 10.1038/Bjc.2011.6.

Rogers, J. M., K. C. Brannen, B. D. Barbee, R. M. Zucker, and S. J. Degitz. 2004. Methanol exposure during gastrulation causes holoprosencephaly, facial dysgenesis, and cervical vertebral malformations in C57BL/6j mice. *Birth Defects Research Part B—Developmental and Reproductive Toxicology* 71 (2):80–88. doi: 10.1002/Bdrb.20003.

Satow, Y., J.Y. Lee, H. Hori, and S. Sawada. 1987. Relative biological effectiveness of Cf-252 with teratogenicity as index. *Annual Report of Research Institute for Radiation Biology and Medicine* 28:133–142.

Schardein, J. L. 1976. *Drugs as Teratogens*. Cleveland, OH: CRC Press, Inc.

Schramm, L., E. Heidbreder, K. Lopau, J. Schaar, D. De Cicco, R. Gotz, and A. Heidland. 1994. Toxic acute renal failure in the rat: Effects of L-arginine and N-methyl-L-arginine on renal function. *Nephrology, Dialysis, Transplantation: Official Publication of the European Dialysis and Transplant Association—European Renal Association* 9 (Suppl 4):88–93.

Shapira, A., Y. D. Livney, H. J. Broxterman, and Y. G. Assaraf. 2011. Nanomedicine for targeted cancer therapy: Towards the overcoming of drug resistance. *Drug Resistance Updates* 14 (3):150–163. doi: 10.1016/j.drup.2011.01.003.

Shukla, S., Z. S. Chen, and S. V. Ambudkar. 2012. Tyrosine kinase inhibitors as modulators of ABC transporter-mediated drug resistance. *Drug Resistance Updates* 15 (1–2):70–80. doi: 10.1016/j.drup.2012.01.005.

Stan, R. V., M. Kubitza, and G. E. Palade. 1999. PV-1 is a component of the fenestral and stomatal diaphragms in fenestrated endothelia. *Proceedings of the National Academy of Sciences of the United States of America* 96 (23):13203–13207.

Stan, R. V., E. Tkachenko, and I. R. Niesman. 2004. PV1 is a key structural component for the formation of the stomatal and fenestral diaphragms. *Molecular Biology of the Cell* 15 (8):3615–3630. doi: 10.1091/mbc.E03-08-0593.

Sullivan, F. M. 1999. Significance of excursions of intake above the acceptable daily intake: Effect of time and dose in developmental toxicology. *Regulatory Toxicology and Pharmacology* 30 (2):S94–S98. doi: 10.1006/rtph.1999.1332.

Tiboni, G. M. and F. Giampietro. 2005. Murine teratology of fluconazole: Evaluation of developmental phase specificity and dose dependence. *Pediatric Research* 58 (1):94–99. doi: 10.1203/01.PDR.0000166754.24957.73.

Tkachenko, E., D. Tse, O. Sideleva, S. J. Deharvengt, M. R. Luciano, Y. Xu, C. L. McGarry et al. 2012. Caveolae, fenestrae and transendothelial channels retain PV1 on the surface of endothelial cells. *PLoS ONE* 7 (3):e32655. doi: 10.1371/journal.pone.0032655.

Tome, L. A., L. Yu, I. de Castro, S. B. Campos, and A. C. Seguro. 1999. Beneficial and harmful effects of L-arginine on renal ischaemia. *Nephrology, Dialysis, Transplantation: Official Publication of the European Dialysis and Transplant Association—European Renal Association* 14 (5):1139–1145.

Walker, C. H. 1978. Species-differences in microsomal mono-oxygenase activity and their relationship to biological half-lives. *Drug Metabolism Reviews* 7 (2):295–323. doi: 10.3109/03602537808993770.

Index